W0254250

Thin Film Coating

Edited by

H. Benkreira
The University of Bradford

The Proceedings of the Second International Symposium on Coating of Thin Films, organised by the Process Technology Group of the Industrial Division of the Royal Society of Chemistry at the Department of Chemical Engineering, University of Bradford, 8–9 September 1992.

Special Publication No. 129

ISBN 0-85186-695-6

A catalogue record for this book is available from the British Library

Published by The Royal Society of Chemistry,
Thomas Graham House, Science Park, Cambridge CB4 4WF

Printed by Hartnolls Ltd., Bodmin

Preface

The articles in this book reflect the proceedings of the second international conference on Coating of Thin Films held at Bradford University (September 1992). The authors are amongst the world leaders in this area and detail many of the advances that have been made in the field of coating flows, rheology, drying and solvent recovery.

Despite the importance of coating technology, it remains a fragmented field of education and research, inevitably leading to a variety of coating methods and approaches to solve the problems arising during operation. This meeting initiates an exchange of views platform and encourages the search for a common, fundamental approach to tackling these problems. Because of the nature of coating processes, rheology, drying and solvent recovery must form an integral part of the study.

This meeting makes a humble beginning to a fascinating project of bringing industry and academia together to study the numerous intriguing coating problems that remain to be solved.

The editor acknowledges the support given by SERC and British industry to research in Coating Technology at Bradford University. The editor also wishes to thank all the participants for supporting this meeting. Gillian Benson deserves a special mention for administrating very ably the symposium here at Bradford.

Contents

Rheology

Rheology

Rheology of Coating Suspensions

D.C-H. Cheng

CONSULTANT IN APPLIED RHEOLOGY, 191 ICKNIELD WAY, LETCHWORTH, HERTFORDSHIRE SG6 4TT, UK

SUMMARY

The viscometric behaviour of a PVC plastisol used in the manufacture of flooring by roll coating is described to illustrate the complex behaviour of coating fluids that are dense suspensions. Such behaviour is due to shear-induced particle migration and leads to viscometric results that depend on the geometry and size of the sensor used and other details of test conditions. This poses a problem in application of the results to predict flow in industrial flow processes, particularly coating processes. To overcome this, it is suggested that viscometers should be used to mimic flow processes. The roll coating of the plastisol is discussed to illustrate what is involved in so doing.

1 INTRODUCTION

Many coating fluids are high solid content suspensions. Such materials are called dense suspensions for short and their rheological properties have been studied by many researchers since the 1930s. I also have been interested in this and my first paper summarising our early work was published in 1978.

It is now well understood that the rheological behaviour of dense suspensions is complicated. I have summarised this in previous publications and some details are given in this paper. The reason is quite simple. It is that dense suspensions are systems of particles suspended in a liquid and they develop non-uniform concentration profiles or structure under shear. Much recent fundamental research has investigated the phenomenon in detail.

Now, conventional interpretation of viscometric results assumes constant concentration and it therefore fails us. When applied to dense suspensions, we do not get the same viscosity values from different viscometers, not just different geometries but also different sizes of the same geometry. This poses a problem for the application of viscosity measurement to predict flow in industrial processes in general and coating processes in particular. The proper interpretation of viscometric results means the separation of the viscosity-concentration function from the concentration profiles in the viscometer. However, even if we can do this, there is still a problem in the application to industrial processes. We need to know the concentration profile in the flow channels and, at present, we do not know how to predict this either.

So, what use is viscometric measurement? This is a problem that I have been trying to answer for several years. The conclusion I have come to is this: that viscometers should be used as mimics of flow processes. I will explain this approach in this paper and illustrate it by describing the viscosity properties of a PVC plastisol, used in the manufacture of foam-backed flooring by roll coating, and consider how the viscosity testing is used to mimic the various unit operations associated with the preparation, handling and coating of the material.

2 RHEOLOGICAL BEHAVIOUR OF A TYPICAL COATING FLUID

The plastisol was a mixture of two PVC polymer powders, with a plasticiser and other ingredients. The volume concentration was 57% based on the dry densities. The actual volume concentration must have been much higher due to swelling of the PVC particles. It was a good example of a dense suspension.

The viscometric testing was made using a Haake RV20 viscometer, with a range of coaxial cylinder and cone-and-plate sensors of different sizes. Cyclic tests were carried out at different ramp rates up to a top shear rate, D_{top}. Both low top shear rate (D_{top} = 10-3 s-1) and high top shear rates (104 s-1) were used. The Newtonian equation was used to calculate shear rate and viscosity.

The viscometric results are described at two levels, as broad-brush results and in fine details. The broad-brush results are the raw data plotted, ignoring details (Fig. 1). At low shear rates, the yield stress varied with gap width of the sensor; and at high shear rates, the flow curves for different gaps converge to 10 Pa.s viscosity. The yield stress varied from 0.5 Pa to 5000 Pa, depending on the gap width (Table 1). The different values obtained with the PK1-1.0 sensor are thought to be an indication of reproducibility connected with sample inhomogeneity. At very high shear rates, the viscosity showed dilatancy (Fig. 2).

In fine detail, the results showed no hysteresis when tested to low D_{top} (Fig. 3). At medium D_{top}, hysteresis loops were obtained (Fig. 4). As D_{top} was taken to higher values, a kink would appear in the flow curve (Fig. 5); and, when tested to very high D_{top}, there would be a catastrophic breakdown in the flow curve (Fig. 6). The kink or onset of the catastrophic breakdown are described as a break point and the associated shear rate varied somewhat with ramp

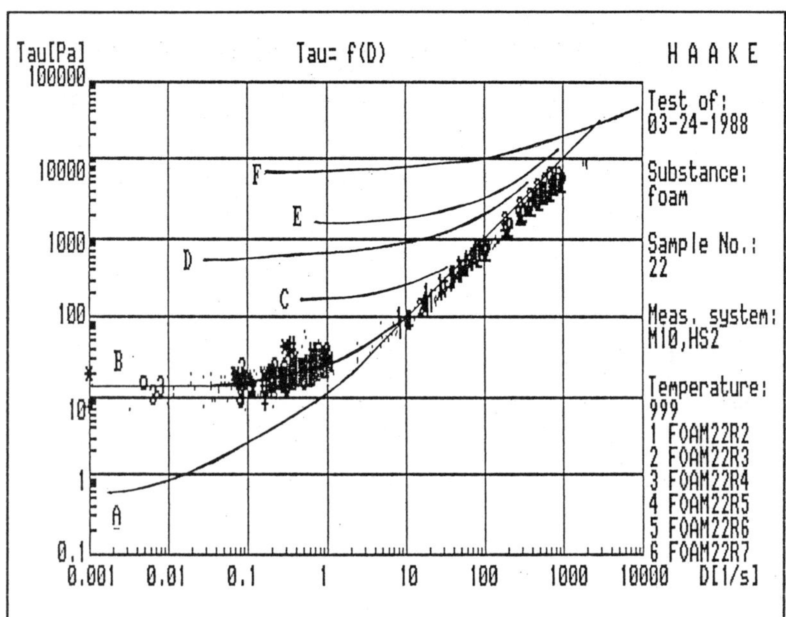

Figure 1 Broad-brush viscometric results for a PVC plastisol measured using different sensors: coaxial cylinder MV3 (A), HS2 (B); cone-and-plate PK1-1.0 (C), PK1-0.3 (D), PK2-0.3 (E), PK3-0.3 (F). See Table 1 for sensor dimensions

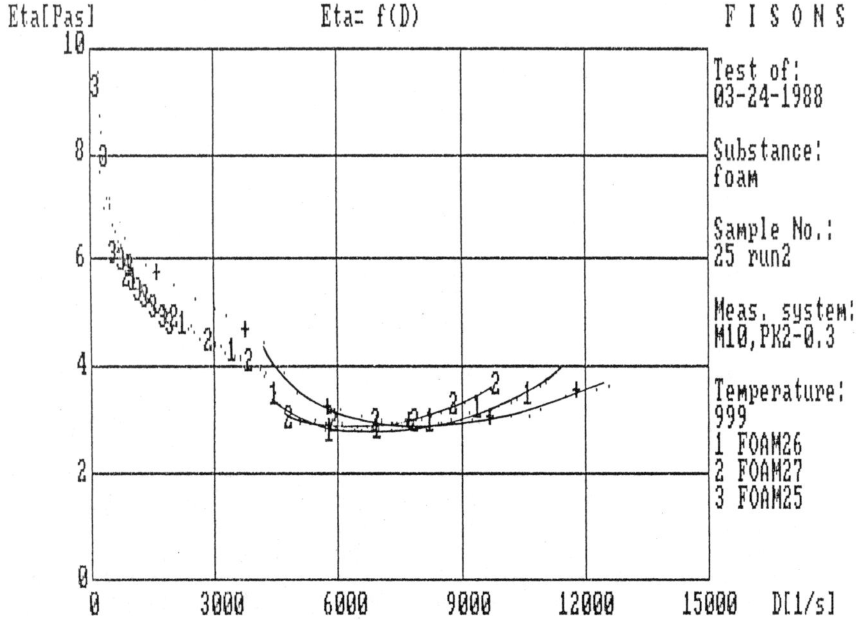

Figure 2 High shear rate result - dilatancy

Table 1. Yield Stress Results

sensor	radius mm	gap width μm	yield stress Pa
coaxial cylinder			
MV3	-	5800	0.5
HS2	-	100	15
cone-and-plate			
PK1-1.0	14	244	100 (22/3) 300 (12/5) 50 (13/5)
PK1-0.3	14	73	500
PK2-0.3	10	52	1500
PK3-0.3	6	31	5000
		gap at edge of cone	(date of test)

rate, but more dramatically with sensor gap (Fig. 7).

Full details of the viscometric testing and results on the PVC plastisol are reported elsewhere[1]. A summary of the complex behaviour is given in Table 2. The significance of the results is discussed in the following section.

3 DENSE SUSPENSION RHEOLOGY

The PVC plastisol described above is just one example of coating fluids. It shows how complex the rheology of these fluids can be. We cannot expect to test each and every different coating fluid as thoroughly as has been done for the plastisol, but, as coating fluids are dense suspensions, we can gain an appreciation of the problems of the rheology of coatings by reference to the understanding of other dense suspensions. There are many other examples in industry and in model systems used in research. I have personally tested many of these since the 1970s. They show the sort of behaviour exhibited by the PVC plastisol, and also other behaviour which is summarised in Table 3. Such behaviour is recognised more and more in recent literature. They have also been described in the earliest pioneering papers, but hidden away in the fine print of the experimental sections[2].

Space limitation does not allow me to describe the different complex behaviour in detail in this paper; this

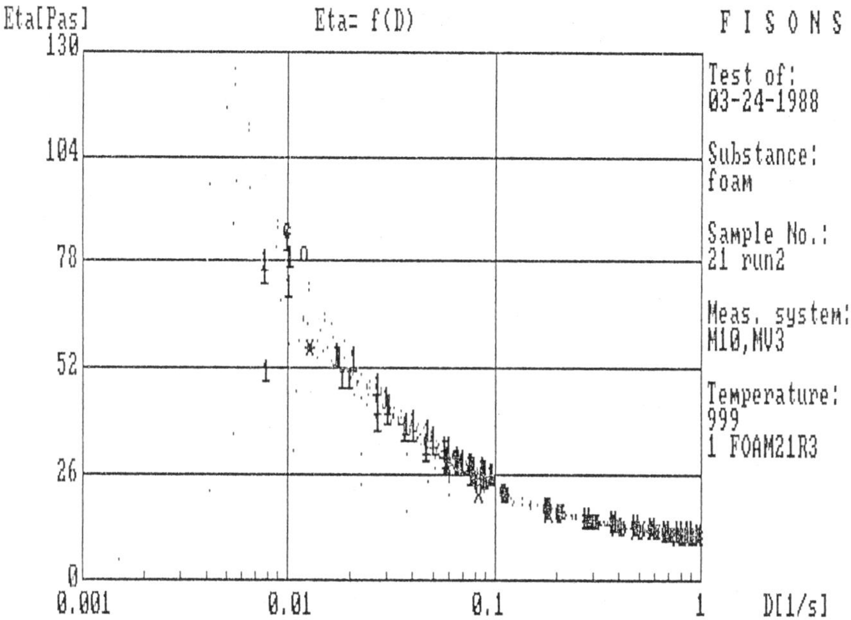

Figure 3 Low shear rate result - no hysteresis

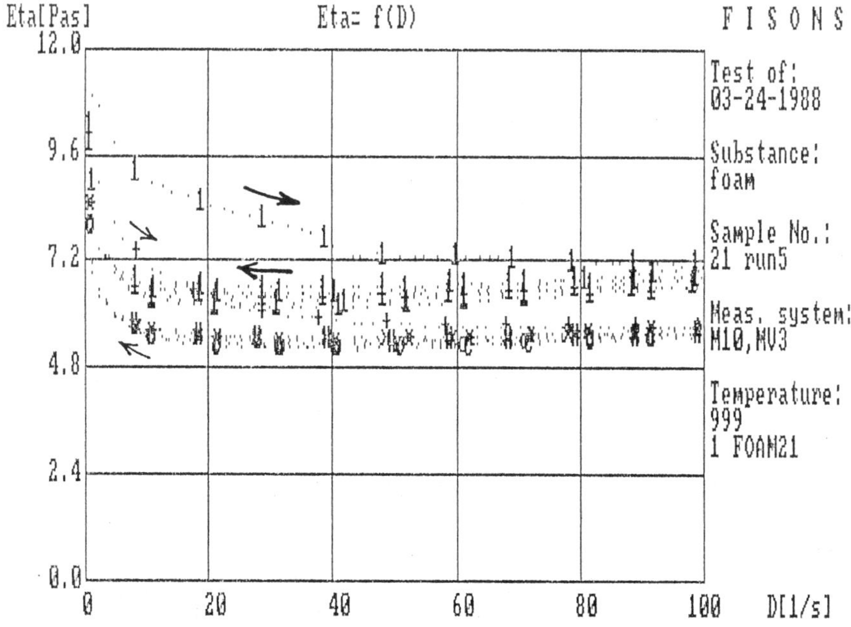

Figure 4 Medium shear rate result - hysteresis loop

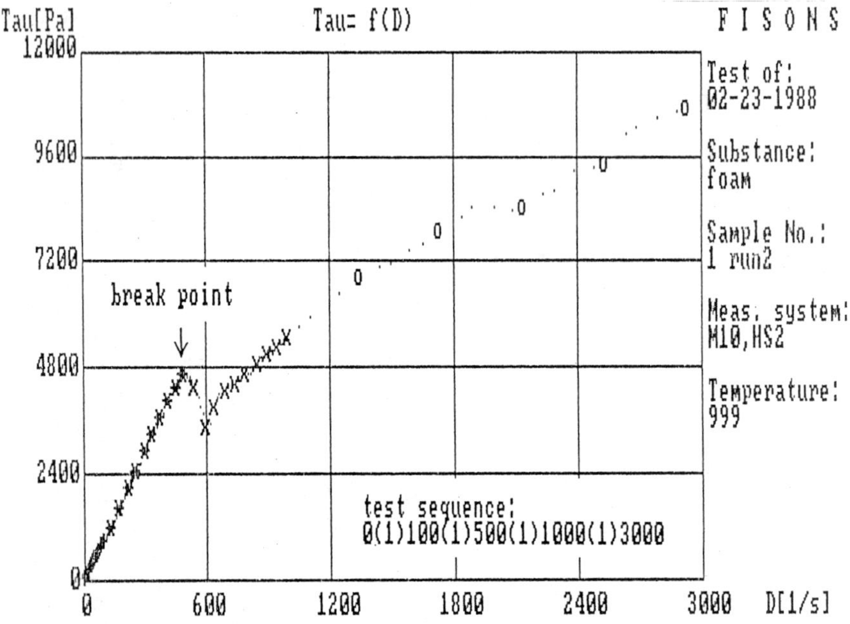

Figure 5 Result showing kink in flow curve

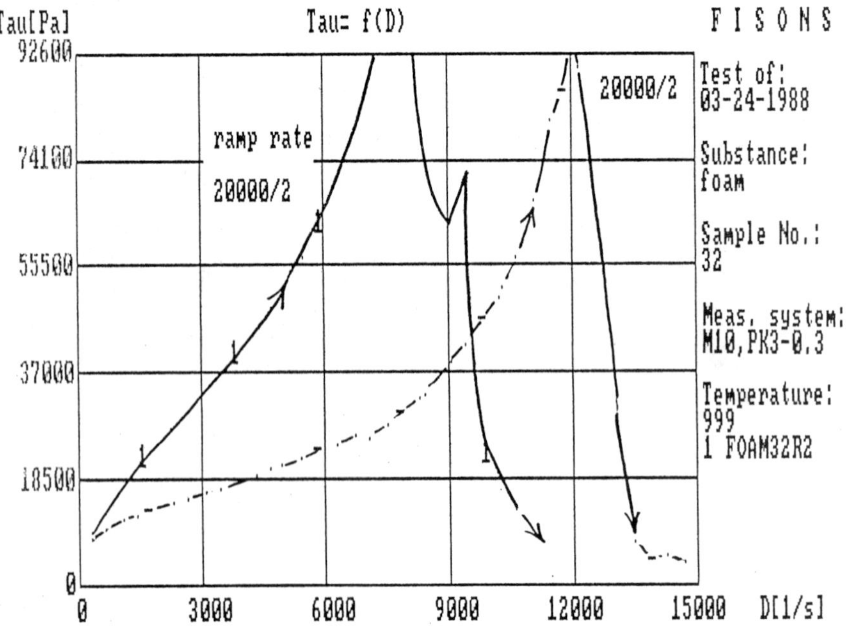

Figure 6 Results showing catastrophic breakdown

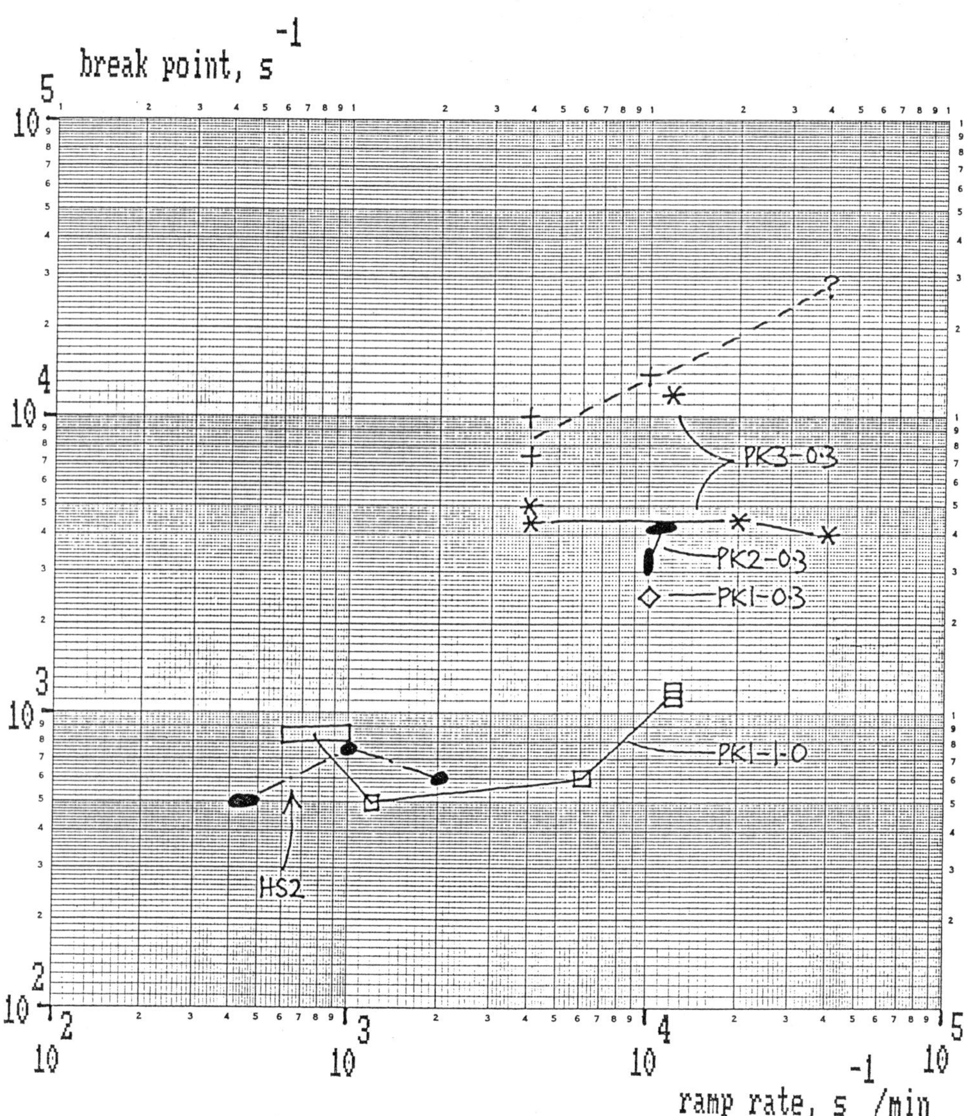

Figure 7 Break point dependence on ramp rate and sensor gap

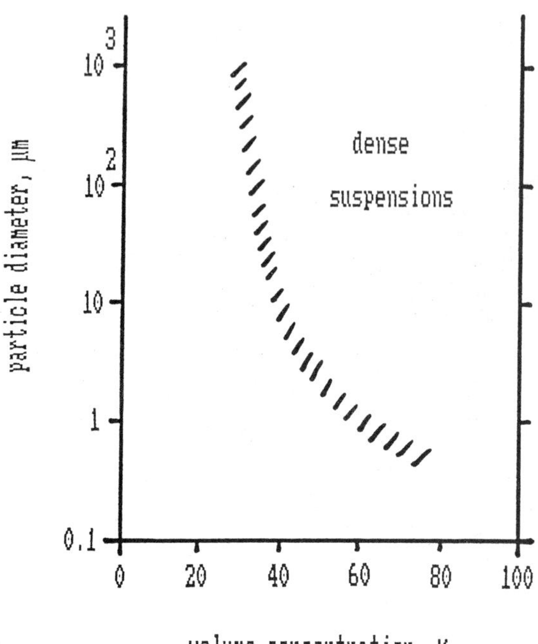

Figure 8 Concentration at which a suspension becomes dense - dependence on particle size

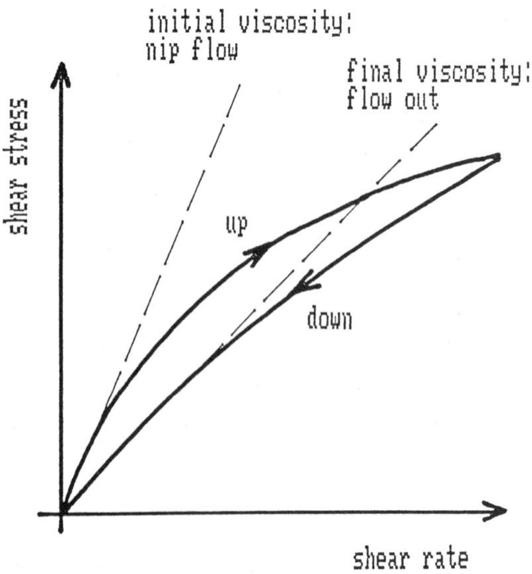

Figure 9 Interpretation of hysteresis loop for roll coating application

Table 2. Summary of PVC Plastisol Viscometric Behaviour

Broad brush

1) low shear rate - yield stress varies with gap width

2) high shear rate - master curve approx. Newtonian

3) very high shear rate - dilatant

Fine details

4) D_{top} << break point - no hysteresis
independent of ramp rate

5) D_{top} < break point - hysteresis loop

6) break point - sudden change in slope of flow curve
or sudden drop in shear stress

7) D_{top} > break point - dilatant
possible second break

8) beyond break - hysteresis loop +ve/-ve
poor repeatability

have been done elsewhere[2-4]. However, it is important to stress that the complex behaviour does not occur only with suspensions of coarse particles, as is often thought, but is found also with fine particles if the solids concentration is sufficiently high. Figure 8 illustrates that the concentration at which a suspension shows complex rheology, and can be described as dense, depends on particle size. Other factors that effect this include particle size distribution, particle shape, shear rate, ramp rate and the viscometer geometry and size.

Previously, some researchers have considered the complicated dense suspension behaviour as artefacts due to poor experimentation. But it is now clear that the reason lies in the multiphase nature of the suspension. They are two-phase mixtures of solid particles in a liquid, which is Newtonian at the very least. Under shear or during flow, particles migrate; there is phase separation, leading to the development of non-uniform concentration or structure in the suspension.

There are many different mechanisms responsible for the particle migration. The phenomenon is not simple and requires mathematics to explain it. In an uniformly unbounded simple shear field, migration is an inertia effect. It does not take place in creeping flow at very low

Table 3. Some Viscometric Behaviour of Dense Suspensions

1) Time dependence

2) Hysteresis loop behaviour

3) Shear reversal behaviour

4) Different responses to different shear histories

5) Oscillatory response or stick-slip under steady shear

6) Sudden breaks in flow curve

7) Drop in viscosity at high concentration

8) Pressure dependence

9) Particle jamming

* Degree of behaviour increases with increasing concentration

* Behaviour varies with dimensions of viscometer sensor

* Behaviour varies with geometry of viscometer sensor

* Results show large scatter that increases with increasing concentration

and zero Reynolds number. The situation is more complicated in non-uniform shear fields, such as wide gap coaxial cylinder or tube viscometers. The presence of a solid wall also encourages migration; this is sometimes described as wall slippage. In concentrated suspensions, jostling of particles would also lead to the development of slip planes or layering of the particles. There is an increasing interest in the literature, theoretical as well as experimental; Table 4 lists some random selections.

4 PROBLEM IN VISCOMETRY AND SUGGESTED SOLUTION

The complex rheological behaviour of dense suspensions, including the PVC plastisol and other coating fluids, is due to the propensity of particles to migrate. This leads to a shear-induced non-uniform concentration profile or structure in a viscometer, whose details depend on sensor geometry and size, and the magnitude of the shear; and it takes time to develop.

The viscometric result depends on the interplay of the viscosity-concentration relationship with the concentration profile during shear. Conventionally, in the interpretation of viscometric results, there is an implied assumption that

Table 4. Some Research into Shear-induced Particle Migration and Micro-structure in Suspensions

Laun et al.[5] (1992)
Rheological and small angle neutron scattering investigation of shear-induced particle structures of concentrated polymer dispersions subjected to plane Poiseuille and Couette flows.

Jomha et al.[6] (1991)
Recent developments in dense suspension rheology.

van der Werff[7] (1990)
The rheology of hard-sphere dispersions.

Li et al.[8] (1988)
Flows of neutrally buoyant suspensions in several viscometers.

Leighton and Acrivos[9] (1987)
The shear-induced migration of particles in concentrated suspensions.

McTique, Givler and Nunziato[10] (1986)
Rheological effects of non-uniform particle distribution in dilute suspensions.

Brady and Bossis[11] (1984)
Dynamic simulation of sheared suspensions.

Husband and Gadala-Maria[12] (1984)
Anisotropy in sheared dilute suspensions of solid spheres.

Nunciato and McTique[13] (1984)
Particle segregation in Poiseuille flow.

Lee[14] (1982)
Aspects of suspension shear flows.

Gadala-Maria and Acrivos[15] (1980)
Shear-induced structure in a concentrated suspension of solid spheres.

(continued)

the concentration is uniform, but this is no longer valid. The governing equations have to be rewritten. For example, with the coaxial cylinder sensor, the concentration (c) depends on radial position (r), as well as dimensions (R_1,R_2) and rotational speed (Ω), and the local viscosity

Table 4. (continued)

Michele, Patzold and Donis[16] (1977)
Alignment and aggregation effects in suspensions of spheres in non-Newtonian media.

Einav and Lee[17] (1973)
Particle migration in laminar boundary layer flow.

Gauthier, Goldsmith and Mason[18] (1971)
Particle motions in non-Newtonian media - I. Couette flow; II. Poiseuille flow.

Highgate and Whorlow[19] (1969)
End effect and particle migration effects in concentric cylinder rheometry.

Cox and Brenner[20] (1968)
The lateral migration of solid particles in Poiseuille flow - I. Theory.

Goldsmith and Mason[21] (1967)
The microrheology of dispersions.

Maude and Yearn[22] (1967)
Particle migrations in suspension flows.

Karnis, Goldsmith and Mason[23] (1966)
The kinetics of flowing dispersions - I. Concentrated suspensions of rigid spheres.

Happel and Brenner[24] (1965)
A sphere in a shearing flow between parallel walls.

Jeffrey and Pearson[25] (1965)
Particle motion in laminar vertical tube flow.

Segre and Silberberg[26] (1963)
Non-Newtonian behaviour of dilute suspensions of macroscopic spheres in a capillary viscometer.

Segre and Silberberg[27] (1962)
Behaviour of macroscopic rigid spheres in Poiseuille flow.

Goldsmith and Mason[28] (1962)
The flow of suspensions through tubes - I. Single spheres, rods, and discs.

varies accordingly:

$$\eta(r) = \eta_o f(c[r, R_1, r_2, \Omega]) \tag{1}$$

where η_o is the viscosity of the suspending liquid. Then:

$$\Omega = \frac{T}{2\pi h}\int_{R_2}^{R_1} \frac{dr}{r^3 \eta(r)}$$

$$= \frac{T}{2\pi h \eta_o}\int_{R_2}^{R_1} \frac{dr}{r^3 f(c[r, R_1, R_2, \Omega])} \tag{2}$$

On using the usual equations for shear stress and shear rate:

$$\tau = \frac{T}{2\pi R_1{}^2 h} \qquad \gamma = \frac{2\Omega}{1-(\frac{R_1}{R_2})^2}$$

the viscosity is:

$$\eta = \frac{1-(\frac{R_1}{R_2})^2}{2R_1^2} \frac{\eta_o}{\int_{R_1}^{R_2} \frac{dr}{r^3 f(c[r, R_1, R_2, \Omega])}} \tag{3}$$

If c is constant, the integral in eq. (3) cancels out $[1-(R_1/R_2)^2]/2R_1{}^2$ and $\eta = \eta_o f(c)$. But in the case of a dense suspension, this is not so and the measured viscosity is seen to depend on the viscometer size, R_1, R_2, and speed, Ω, or shear rate.

In the case of the tube viscometer, there is a different concentration profile, c(r,R,V), to contend with:

$$\eta(r) = \eta_o f(c[r, R, V]) \tag{4}$$

The velocity is then:

$$V = \frac{P}{2\pi L R^2 \eta_o}\int_0^R \frac{r^3 dr}{f(c[r, R, V])} \tag{5}$$

and the measured viscosity:

$$\eta = \frac{\frac{R}{2}\frac{P}{L}}{\frac{4V}{R}} = \frac{R^4}{4} \frac{\eta_o}{\int_0^R \frac{r^3 dr}{f(c[r,R,V])}} \tag{6}$$

Again, if c is constant, $\eta = \eta_o f(c)$. But, in general, the measured viscosity depends on the tube radius, R, and the velocity, V, or shear rate.

Because the two concentration profiles, $c[r,R_1,R_2,\Omega]$ and $c[r,R,V]$, are different, the measured viscosity from the coaxial cylinder sensor would be different from the tube viscometer. This means that, in general, viscosity results calculated using the conventional equations would depend on the viscometer geometry and the sizes of the actual sensor used.

To interpret viscometric results properly, we have to separate the two factors, concentration profile from the viscosity-concentration relationship. However, current knowledge does not allow us to do that.

In any case, to apply the viscosity-concentration relationship (assuming that we can isolate this) to predict flow, we need to know the concentration profile in the flow process. For example, if we wish to use coaxial cylinder viscometer data to predict pipeline pressure drop, we need to be able to predict c(r,R,V) also. As we do not know how to do this, what use is viscometric measurement? Clearly, research is indicated, and this is indeed in progress in academia, as the references in Table 4 testify. In the meantime, what should the applied rheologist do?

We could perhaps follow engineering practice and resort to the scale-up approach, and abandon viscometry for pilot-scale experiments. But this is expensive and time consuming so I propose a new approach, namely to use viscometers to mimic flow processes[2, 29]. By this I mean that the conditions of testing should closely match those of the application flow situation, so that what happens to the particles in the flow process should be reproduced in the viscometer. In this way, the results of viscometric tests may be relevant to the application situation.

In general terms, this means that the choice of viscometer sensor should closely match the flow process in terms of geometry and dimensions (or size), and the shear history of the sample in the viscometer should be as close as possible to that in the process. Depending on which feature of the process is important, the sensor gap width should be as narrow as the narrowest gap or as wide as the widest channel. In determining the shear history, the lowest and highest shear rates in the process have to be estimated, and also the residence time, whether the passage is swift or whether there are dead areas in which the fluid is held for long periods. The trapped fluid escapes at irregular intervals and may cause problems downstream.

It is difficult to generalise and lay down specific guidelines for the implementation of this recommendation. So far in writing and talking about this[2, 29] I have discussed the matter only in general terms. In the rest of this paper, I will discuss how it may be applied in the preparation, handling and roll coating of the PVC plastisol in the manufacture of the flooring.

5 APPLICATION TO PROCESSING OF THE PVC PLASTISOL

The viscometric behaviour of the PVC plastisol has already been described earlier. In this section, I discuss how the results may be applied in the preparation, handling and roll coating of the plastisol in flooring manufacture[1]. There are many objectives to be met at different stages of the manufacture and each one has to be considered individually.

The first is the preparation of the plastisol by mixing. It must not be so viscous that it overloads the mixer motor and adequate mixing is not achieved. It is pumped, and again, the viscosity must not be too high or the pumping pressure would be excessive and it would not be conveyed at the desired throughput. There are in fact design equations that can be used to predict the mixer power requirement and the pipeline pressure drop using the viscometric flow curve. In these unit operations, the plastisol is handled in bulk and the flow channels are wide. The correct flow curve to use is that determined using sensors with as wide a gap as possible. To increase one's confidence in the validity of the viscometric results, different large gap widths should be used to check how sensitive the results are to the width.

The plastisol is filtered on vibrating screens. It would not pass through the screen if the viscosity is high. Although not mentioned in the early part of this paper, it is known that the shear viscosity of dense suspensions is affected by superimposed oscillatory shear[30]. This means that conventional steady viscometry may not give the relevant results. The plastisol has to be tested under combined steady and oscillatory shear. The frequency and amplitude of the oscillation have to be the same as on the screen, and the viscometric gap has to be similar to the screen aperture.

The most important aspect of the plastisol behaviour is found in the roll coating process. The choice of the viscometer sensor and test procedure follow from the knowledge of the plastisol rheology taken together with the requirements of the coating process. There are three factors to consider: gap width, shear rate and time of shear.

Ideally, the viscometer gap should be the same as the gap between the roller and the substrate and a narrow gap sensor should be chosen. The flooring company had previously decided on the Haake instrument and in the QC laboratory, the HS sensor (gap 100 μm) had therefore been chosen to match the gap on the roll coater.

On the other hand, after deposition on the substrate, the plastisol undergoes flow-out. The coating film is not confined in this situation and the viscometer gap should therefore be wider to correspond to the film thickness. The use of the one-degree PK1-1.0 cone (with 244 μm gap), or even the MV3 sensor (5.8 mm gap), would be more relevant.

The usual accepted estimate of shear rate in the roller nip puts it at 10^6 s^{-1} and conventional wisdom advises that one should attempt viscometric testing to this high shear rate. But, if a thick coat is being deposited, a lower shear rate would apply. Assuming a 1 mm film and a substrate speed of 60 m/min, the shear rate would be only 1000 s^{-1}.

In other parts of the coating machine, lower shear rates still are involved. For example, some estimate for the draw-up flow around the pick-up roll puts the shear rate at 10-50 s^{-1}. Recirculation in the plastisol reservoir tank would be at an even lower shear rate, maybe only 1 s^{-1}.

When the shear rate associated with the levelling or the flow out of the film on the substrate is considered, shear rates of 0.1 or 0.01 s^{-1} have been quoted.

The time of shear suffered by the plastisol in the roll coating process also varied widely. In many parts of the process, the plastisol undergoes prolonged shear, such as in the reservoir tank holding the bulk fluid. The coating fluid is gradually drawn from the reservoir tank (or from the recirculating pool) and it will have a long shear history before it is deposited onto the substrate.

In sharp contrast, the residence time of the plastisol in the nip is only a fraction of a second. Again, it is seen that there is a range of time of shear to contend with.

Table 5 is a summary of the gap widths, shear rates and times of shear encountered in the roll coater. It is clear that it would not be adequate to assess the viscosity of the plastisol at just one set of test conditions. Furthermore, it is seen that the conditions are correlated. The typical element of plastisol would be prolongedly sheared at a low shear rate in the wide gap reservoir, followed by a short shear at high shear rate through the narrow nip, and then a longish shear at a very low shear rate during flow-out. In the viscometer, the testing shear history should reflect this shear history. It is of course not practical to vary the sensor gap during a test, but at least a test procedure involving long shear at a low shear rate, then short shear at high shear rate, and then followed by long shear at very low shear rate can be attempted.

Table 5. Matching Viscometer Test Conditions to Roll Coating Process

	fluid pond or reservoir pick-up	roller nip	film on substrate
1) gap width	wide	narrow	medium
2) shear rate, s^{-1}	1	10^3-10^6	0.01-0.1
3) time of shear	prolonged	<< 1 s	long

Some simplification of the testing may be possible. This happens if one can identify a particular stage of the coating process or shear history as the main problem. Then the appropriate gap, shear rate and time of shear would be used in the viscometric testing.

A fourth factor to consider is that shear-induced changes in the plastisol gives rise to hysteresis loops in cyclic shear (Fig. 9). One way to interpret the loop result is to consider that the initial higher viscosity of the up-curve relates to the flow of relatively fresh fluid through the nip, while the lower viscosity of the down-curve would be applicable to levelling during flow-out. Therefore different viscosity values should be taken from the hysteresis loop in order to explain any problem in the nip flow and during flow-out. However, one should be careful in choosing the cyclic shear conditions. The sample should be pre-conditioned first to simulate the reservoir shear history. And the choice of gap, shear rate and ramp time has to be carefully considered as already discussed.

At high shear rates, the plastisol undergoes catastrophic break-down, which is exasperated with prolonged shear. This leads to chaotic and uncontrolled behaviour. This will lead to corresponding defects in the coating film, and is clearly to be avoided in the coating process. If this were to happen, it would happen in the nip. The shear rate levels in the nip are known to some degree; we cannot really reduce this or the production throughput would not be economic. We have to avoid it through QC of the plastisol viscosity properties. In formulating the plastisol, a requirement is to have as low a plasticiser content as possible, without leading to break-down. In viscometric testing, one should therefore push the samples beyond the break point in order to determine the limiting compositions of different formulations to ensure trouble free operation of the coating process. Usually, when testing dense suspensions, people try to look for viscometric testing conditions that give nice smooth flow curves. I am advocating just the reverse. Avoiding trouble in viscometers may give you trouble in the process. Therefore, look for trouble in your viscometer in order to avoid trouble in the coating process.

6 EXTENSION TO OTHER COATING PROCESSES

The description of the plastisol roll coating process in the preceding section shows how one must analyse the process in order that the viscometric testing can give relevant results. This has many counter parts in other coating processes, such as paper coating, transfer printing and screen printing. I do not propose to discuss these in detail, but there are some interesting variations.

For example, in screen printing, a reciprocating squeegee is used, both to pre-condition and to spread the ink. To mimic this in the viscometer, the sweep time in the cyclic shear should be of the order of seconds, compared with fraction of a second when mimicking the flow through the roll coater nip.

Another variation is that paper coating and printing involves porous and absorbent substrates. With the penetration of the liquid medium into the substrate, the solids concentration of the film would become higher than that in the bulk fluid. In viscometric testing, one should investigate the effect of concentration higher than the bulk value used in normal production. ---- To take the idea of mimicking further, one could wrap the viscometer sensor with paper before testing a sample.

7 CONCLUDING REMARKS

In discussing the rheology of suspensions, it is usual to consider either the phenomenological aspects of viscometric behaviour and/or the physical and chemical bases of material behaviour. What I have tried to show in this paper is that the rheology of dense suspensions depends very much on the viscometric testing conditions and the details of the flow in industrial processes. The reason for this complex rheology, in simple terms, is that dense suspensions are multiphase systems in which the concentration distribution or internal structure is profoundly affected by shear and flow. Conventional viscometric measurement, which assumes uniform concentration, therefore gives viscosity results that vary with sensor geometry and sizes. This makes it impossible, at the current state of understanding, to use the results to predict behaviour in process flows. This raises the question of how one may use viscometers to give meaningful results. It is suggested that viscometers should be used to mimic flow processes. It means, in general terms that the viscometric test conditions should be matched as closely as possible to the flow process, in respect of flow geometry, channel dimensions and shear rate-time history. Because different conditions exist in different parts of the process, viscometric testing may involve the exploration of the suspension behaviour using a variety of conditions. If some particular aspects of the process can be identified as of major importance, then it may be possible to restrict the test conditions accordingly.

Many coating suspensions are dense suspensions and the PVC plastisol used in the manufacture of flooring by the roll coating process is an example. This paper describes its viscometric behaviour to illustrate the complex rheology of dense suspensions and discusses what it entails when using the viscometer to mimic the roll coating process. Other coating processes are also briefly discussed. It is not possible to be more specific in laying down guidelines on how to use viscometers to mimic coating processes. It is hoped that the example of the PVC plastisol roll coating gives the reader a feel of what is involved.

ACKNOWLEDGEMENT

This paper is based on work done at Warren Spring Laboratory, Stevenage. The Laboratory's permission for publication is gratefully acknowledged.

REFERENCES

1. D C-H Cheng, 'Viscosity measurement on a PVC plastisol as an example of dense suspensions'. Rept No. LR 711 (MP/BM). Stevenage: Warren Spring Laboratory, 1989.
2. D C-H Cheng, 'Flow properties of suspensions of inert spheres'. Rept No. LR 548 (MH). Stevenage: Warren Spring Laboratory, 1984.
3. D C-H Cheng and R A Richmond, Rheol Acta, 1978, 17, 446-453.
4. D C-H Cheng, Powder Tech, 1984, 37, 255-273; Brit Soc Rheol Bull, 1984, 27, 1-8.
5. H M Laun et al., J Rheo, 1992, 36, 743-787.
6. A I Jomha et al., Powder Tech, 1991, 65, 343-370.
7. J C van der Werff, Ph D Thesis, University of Utrecht, 1990.
8. A C Li et al., Chem Eng Comm, 1988, 73, 95-119.
9. D Leighton and A Acrivos, J Fluid Mech, 1987, 181, 415-439.
10. D F McTique, R C Givler and J W Nunciato, J Rheol, 1986, 30, 1053-1076.
11. J F Brady and G Bossis, in 'Proc IXth Intl Congr Rheol, Vol. 2. Fluids', pp. 475-480. Mexico: Univ Nacion Auton Mexico, 1984.
12. D M Husband and F Gadala-Maria, in 'Proc IXth Intl Congr Rheol, Vol. 2. Fluids', pp. 521-528. Mexico: Univ Nacion Auton Mexico, 1984.
13. J W Nunciato and D F McTique in 'Proc IXth Intl Congr Rheol, Vol. 1. Theory', pp. 345-352. Mexico: Univ Nacion Auton Mexico, 1984.
14. S L Lee, Adv Appl Mech, 1982, 22, 1-65.
15. F Gadala-Maria and A Acrivos, J Rheol, 1980, 24, 799-814.
16. J Michele, R Patzold and R Donis, Rheol Acta, 1977, 16,

317-321.

17. S Einav and S L Lee, Int J Multiphase Flow, 1973, 1, 73-88.
18. F Gauthier, H L Goldsmith and S G Mason, Rheol Acta, 1971, 10, 344-364; Trans Soc Rheol, 9171, 15, 297-330.
19. D J Highgate and R W Whorlow, Rheol Acta, 1969, 8, 142-151.
20. R G Cox and H Brenner, Chem Eng Sci, 1968, 23, 147-173.
21. H L Goldsmith and S G Mason, in F R Eirich (Ed), 'Rheology: Theory and Applications', Vol. 4, Chap.2, pp.85-250. New York: Academic Press, 1967.
22. A D Maude and J A Yearn, J Fluid Mech, 1967, 30, 601-621.
23. A Karnis, H L Goldsmith and S G Mason, J Coll Interf Sci, 1966, 22, 531-553.
24. J Happel and H Brenner, 'Low Reynolds Number Hydrodynamics with Special Applications to Particulate Media', pp. 328-329. Englewood Cliffs, NJ: Prentice Hall, 1965.
25. R C Jeffrey and J R A Pearson, J Fluid Mech, 1965, 22, 721-735.
26. S Segre and A Silberberg, J Coll Sci, 1963, 18, 312-317.
27. S Segre and A Silberberg, J Fluid Mech, 1962, 14, 115-157.
28. H L Goldsmith and S G Mason, J Coll Sci, 1962, 17, 448-476.
29. D C-H Cheng, in D R Oliver (Ed), 'Third European Rheology Conference', pp.94-97. London: Elsevier Applied Science, 1990.
30. A I Jomha and L V Woodcock, Trans IChemE, 1990, 68A, 550-557.

The Use of Rheological Techniques in the Study of Interactions Between Components of Paper Coating Colours

D.J. Preston, P.M. McGenity, and J.C. Husband
ECC INTERNATIONAL LTD., JOHN KEAY HOUSE, ST. AUSTELL, CORNWALL PL25 4DJ, UK

1. ABSTRACT

Oscillatory and steady state (rotational) rheological techniques are useful in the study of interactions between components of paper coating colours containing kaolin pigments.

In offset formulations, soluble polymers such as sodium carboxymethyl cellulose (CMC) are frequently used as thickeners. Forced oscillation measurements show that the effect of the polymer is to increase the elastic response of the colour at low strains. This is shown to be a result of an interaction between the clay and the soluble polymer.

These weakly flocculated structures are shear sensitive, and the shear-thinning behaviour observed can be predicted as a function of shear rate using the Sisko model, a modified power-law relationship. Both the degree of shear-thinning and the elasticity of the colours were greatly increased by relatively low levels of CMC addition.

2. INTRODUCTION

The coating of paper is carried out for a number of reasons. For example, properties such as gloss, smoothness and opacity are greatly enhanced. The quality demands of the printing process usually require a coated sheet. A useful introduction to paper coating has been published by Hagemeyer[1]. The most commonly used coating methods involve either passing the uncoated paper through a "pond" of coating suspension, or applying a film of coating using a roll applicator. Whatever method is used, it is necessary to remove the excess coating, usually with a metering blade. Typically, coat weights of between 5 and 10 $g.m^{-2}$ are applied to both sides of the paper, representing an average wet film thickness of the order of 5 µm. Many modern paper coating machines operate at high speeds (1000 - 1450 $m.min^{-1}$) and shear rates under the blade have been estimated[2] at over 10^6 s^{-1}. This places heavy demands upon the pigment and the formulation in which it is used. In particular, the rheology of the coating formulation (usually termed the **coating colour**) is important and can be influenced not only by the constituents present but also by any interactions between them.

Paper coating colours are high-solids colloidal dispersions (up to 45% by volume) prepared by mixing a pigment (usually kaolin or calcium carbonate or a blend of the two) in a deflocculated condition with a binder or mixture of binders. Commonly used binders are polymer latices based on styrene and butadiene or acrylic monomers. In many instances it is also necessary to add a flow-modifier which causes low-shear thickening of the colour, conferring shear-thinning or pseudoplastic behaviour. The flow-modifier also controls the rate of water loss into the base paper during the coating application. Typical flow-modifiers are water-soluble polymers such as sodium carboxymethyl cellulose or carboxylated acrylic copolymers.

Such a combination of dispersed phases and soluble polymers displays complex interactions which can be studied using rheological techniques. In addition to conventional rotational viscometry, more recent work has used an oscillatory shear technique which is capable of discriminating between the elastic and viscous elements contributing to suspension rheology.

3. EXPERIMENTAL

Materials

A fine coating clay from Georgia, U.S.A., was employed in most of the experiments described here. 94% by weight of particles were finer than 2 μm e.s.d. Suspensions were prepared using 0.3 wt.% sodium polyacrylate as the dispersant. The following offset formulation was used :

100 parts clay
11 wt.% latex
0 - 1 wt.% sodium carboxymethyl cellulose (CMC).
pH ca. 8.0

The latex was a commercial carboxylated styrene-butadiene copolymer of mean particle size 0.18 μm and a glass transition temperature of 20°C (measured by Dynamic Mechanical Thermal Analysis[3]). A low viscosity grade of CMC having a molecular weight of 34,000 (determined from dilute solution viscometry[4]) and a degree of substitution (DS) of 0.7 was used.

Rheological Measurements

Measurements of rheological parameters were carried out using a Bohlin VOR controlled strain rate rheometer. Viscosity versus shear rate data were collected between 58 s^{-1} and 1470 s^{-1} using stepwise increases in shear rate with a shear time of 20 s at each step. Forced oscillation experiments were conducted by carrying out a frequency sweep at a fixed amplitude, 2.0 mrad. Previous experiments demonstrated that this was within the region of linear viscoelastic response for this type of coating colour. A useful review of the principles of oscillatory testing has been published by Goodwin[5].

4. RESULTS

Rheological Effects of CMC Addition

Rotational Shear. Figure 1 shows the effect on viscosity of adding CMC to the clay slurry in the absence of latex, at a constant solids level of 64.0 wt%. A linear increase in viscosity is obtained as the dose is increased up to 1.0 wt.%, towards the upper limit of commercial use. Also shown in Figure 1 is the theoretical viscosity calculated from the viscosity of the CMC-free suspension and the relative viscosity of the CMC solution, making the assumption that the only effect of CMC is to viscosify the aqueous phase.

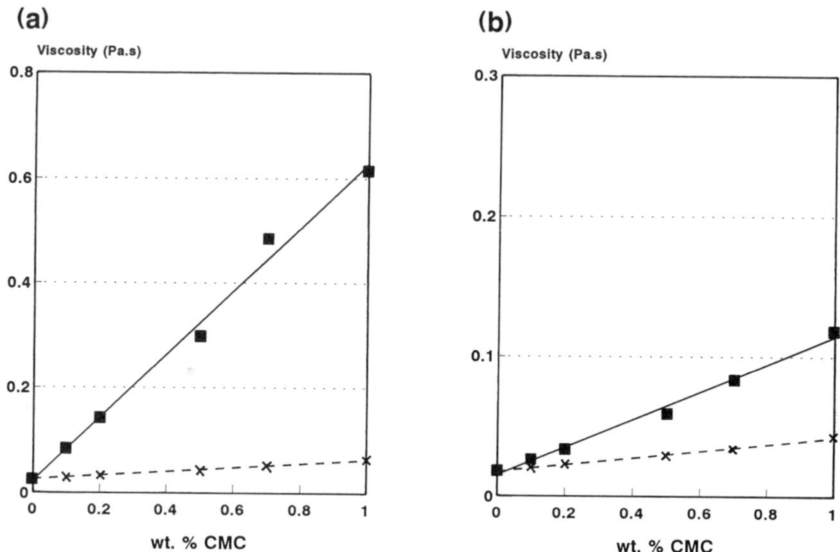

Figure 1. Viscosity of clay suspensions at increasing CMC doses, 64 wt.% solids, (a) at 58 s^{-1}, (b) at 1470 s^{-1}. Solid line : experimental points, dotted line : calculated values.

The calculation will predict the viscosity correctly only in the absence of interactions between the components. Clearly, there is some interaction since the measured viscosity values are much higher than the calculation predicts. The measured values are closer to the calculated ones at the higher shear rate, suggesting that the interaction is most powerful at low rates of shear.

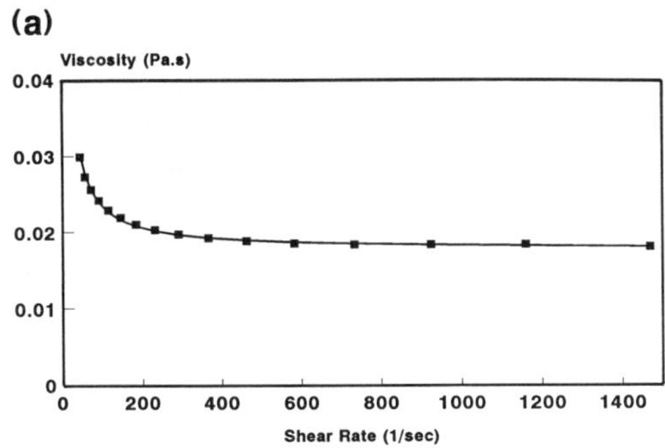

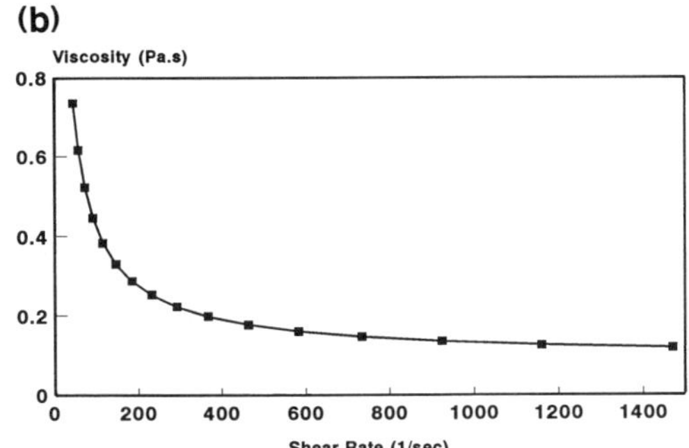

Figure 2. Viscosity vs. shear rate data for clay suspensions containing (a) no CMC, (b) 1.0 wt% CMC (64 wt% solids). Points are experimentally determined; the solid line is a computer fit using the Sisko model.

The behaviour of shear-thinning suspensions such as these can be modelled using a modified "power-law" relationship such as the one proposed by Sisko[6]:

$$\eta = \eta_{lim} + K\ G^{n} \quad \text{.....................[1]}$$

where η = suspension viscosity

η_{lim} = limiting viscosity at high shear

G = shear rate

This equation has been particularly successful in modelling the behaviour of a range of industrial suspensions from lubricating grease to yoghurt[7]. The results for clay suspensions before and after CMC addition are shown in Figure 2. An excellent fit was obtained to the measured data at all levels of CMC addition.

The results illustrated in Figures 1 and 2 suggest that the suspensions become more sensitive to shear as the CMC dose increases. In order to quantify the degree of shear-thinning when a good fit to the model is obtained, equation [1] can be differentiated with respect to shear rate as follows :

$$d\eta/dG = n\ K\ G^{n-1} \quad ....[2]$$

If at any given rate of shear, $d\eta/dG$ is negative, then the system is shear-thinning at that point. Furthermore, the larger the negative value of $d\eta/dG$, the more strongly shear-thinning the system is. Results for a shear rate of 100 s^{-1} are shown in Figure 3. The plot confirms a more powerful shear-thinning effect with increasing levels of CMC, with the most pronounced change occurring after the addition of the first 0.2 wt.% increment.

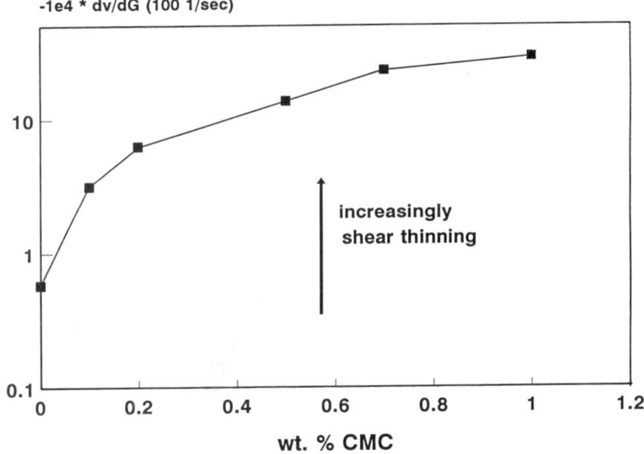

Figure 3. Shear thinning of clay suspensions at 100 s^{-1} as a function of CMC dose (64.0 wt.% solids).

Oscillatory Shear. An oscillation study was also conducted on these suspensions. The addition of CMC increased the elastic response, as shown by a lowering of the phase angle from 25° to 7° (Figure 4). Once again, it is striking that the most pronounced change occurs after the first 0.2 pph of CMC has been added, and suggests a link between elasticity and the degree of shear-thinning of the suspensions. That the elasticity is not a result of the polymeric nature of CMC itself was demonstrated in an oscillation experiment using 11 wt.% CMC solution (8 times more concentrated than would be found in the aqueous phase of the suspension containing 1.0 wt.% in Figure 4). The phase angle was 85°, indicating that the solution has virtually no elastic character (Figure 5).

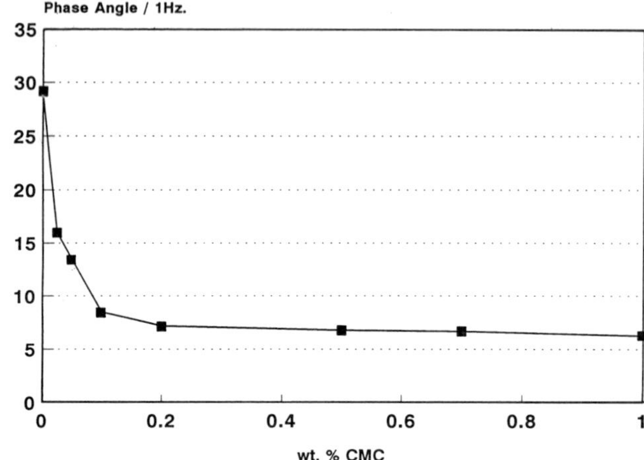

Figure 4. Forced oscillation of clay suspensions : phase angle vs. CMC dose, 64.0 wt.% solids, 1 Hz.

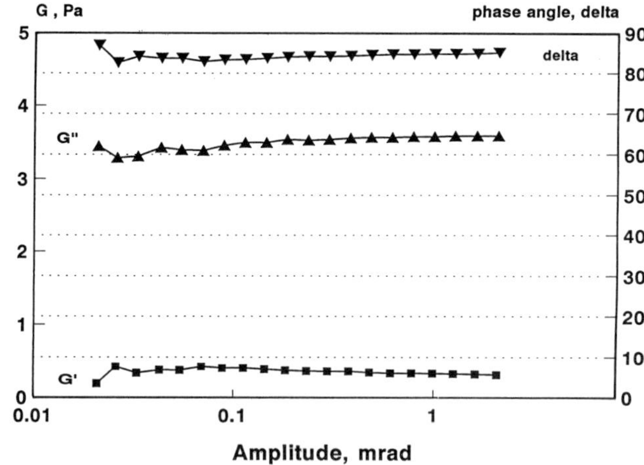

Figure 5. Forced oscillation, 11.0 wt% CMC solution, 1 Hz. (▼) phase angle δ ; (▲) storage modulus G' ; (■) loss modulus G".

This increase in the elasticity of coating colours and clay suspensions has been reported by several workers[8,9,10,11] and is evidence of an interaction between CMC and the pigment which leads to the formation of a floc structure. Similar rheological effects have been observed in the presence of soluble acrylic thickeners[12]. Such a structure is highly elastic at the low rates of strain involved in an oscillation experiment and the shear-thinning behaviour observed suggests that it is destroyed at moderate rates of shear.

Interactions between CMC and clay

Previously published data[11,13] has reported that CMC is adsorbed onto kaolinite and that the effect is competitive with polyacrylate dispersant. The adsorption of CMC onto the clay used here (pretreated with polyacrylate) was measured at 40 wt.% suspension solids at 21°C. The equilibrium CMC concentration was estimated using the anthrone method[14]. The isotherm obtained is shown in Figure 6 and indicates adsorption of CMC up to a plateau level of 2.0 mg.g^{-1}. This is of a similar order of magnitude to previously published results for clay in the presence of polyacrylate.

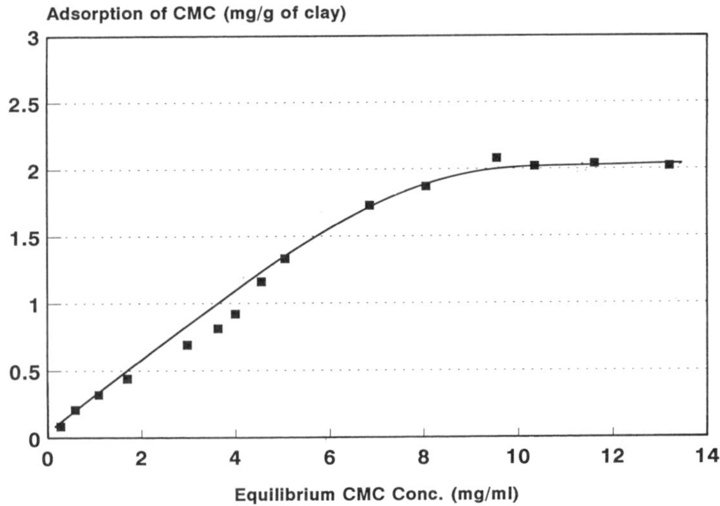

Figure 6. Adsorption of CMC onto clay, 40 wt% solids; 21°C.

The aggregation of clay particles by CMC was demonstrated using Photon Correlation Spectroscopy (PCS)[15]. To avoid problems of polydispersity, a specially-prepared narrow size fraction of an English clay was used. The samples were diluted for PCS using CMC solution of varying concentrations calculated to maintain the [CMC] in the aqueous phase of the suspensions under study. Appropriate viscosity corrections were made[16]. Figure 7 shows mean clay particle size as a function of [CMC] in the aqueous phase. Evidently flocculation is occurring, with a maximum at a [CMC] equivalent to a dose of 0.2 wt.%. Although possibly fortuitous, this dose coincides with the rheological data in Figures 3 and 4.

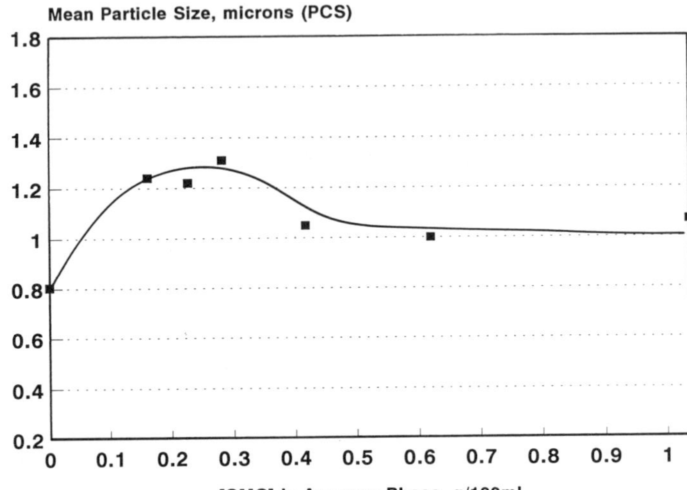

Figure 7. Flocculation of 0.6 - 0.7 μm kaolin by CMC.

Soluble polymers are known to induce flocculation by two possible mechanisms : bridging[17] and depletion, or exclusion, flocculation[18,19]. (Charge neutralisation can be ruled out in this instance since clay and CMC both carry an anionic charge). Currently we prefer to interpret these results in terms of depletion flocculation since interparticle bridging by adsorbed polymer of low molecular weight seems unlikely. In our previous study of interactions between CMC and latex (similar to those employed here) the evidence was strongly in favour of a depletion mechanism since the lowest molecular weight CMC had the greatest effect[11]. Such weakly-flocculated suspensions are reported to be highly elastic[18] and strongly shear-thinning[19].

Rheological study of clay/latex/CMC system

Oscillatory shear experiments were also performed on a series of "colours" containing clay,latex and CMC or combinations thereof. When considered on a volume solids basis, the effects of interactions between components can be directly compared. Storage modulus and phase angle results are plotted in Figure 8.

The addition of latex to the clay slurry raised the storage modulus slightly but the overall effect on phase angle was negligible. Fadat et al[20] reported an interaction between a carboxylated latex and clay which resulted in the deposition of a small proportion of latex onto the pigment. Our results, however, suggest that if such an interaction is taking place, it does not significantly alter the structure of the suspension. Adding 1.0 wt.% CMC resulted in a 100-fold increase in the storage modulus and a marked decrease in the phase angle. The phase angle results suggest that the presence of latex

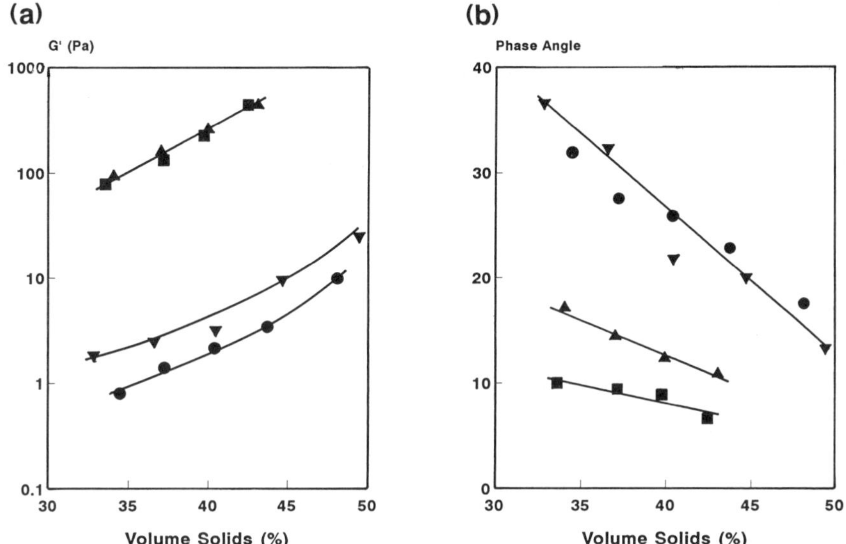

Figure 8. Forced oscillation of coating colours at 5 Hz. (a) Storage modulus, G', (b) Phase angle, δ. (●) clay only; (▼) clay/latex; (■) clay/CMC; (▲) clay/latex/CMC.

slightly reduces the elasticity of the clay suspension after addition of CMC.

5. CONCLUSIONS

The use of rheological techniques, especially rotational and oscillatory shear methods, is a powerful tool in the study of interactions between components of paper coating colours.

Sodium carboxymethyl cellulose is a frequently used thickener in European offset formulations. Rheological studies have shown that the mode of thickening under very low shear conditions (such as in oscillation experiments) is different than at higher rates of shear (1500 s^{-1}).

A highly elastic floc structure is formed under low shear conditions. This is primarily a result of an interaction between CMC and the clay. As the rate of shear increases, the structure is progressively broken down until the major influence upon suspension viscosity is the viscosity of the aqueous phase. The effects of this interaction on the paper coating process is still the subject of much speculation. Whilst the shear rates involved in blade coating are undoubtedly several orders of magnitude larger than the highest used in our experiments, the deformation times are extremely short[21]. It is therefore possible that some floc structure is able to survive the blade. Some evidence for this may be obtained from a consideration of the effects of CMC on high-speed runnability and sheet properties. McGenity et al[11] and Gane et al[22] have shown that the presence of soluble polymers such as CMC in a formulation can encourage the formation of blade deposits, spits and streaks. Such phenomena have been linked to the presence of viscoelasticity in the colour[23,24]. In order to shed more light on these problems, future work needs to address the measurement of relaxation times of coating colours at appropriate shear rates.

Acknowledgements

The authors wish to thank Professor Ken Walters for helpful discussions and the board of directors of ECC International for permission to publish this work.

REFERENCES

1. R.W.Hagemeyer, "Pigment Coating", in "Pulp and Paper Chemistry and Technology Vol.IV", ed. J.P.Casey, Wiley Interscience, New York, 1983, p.2013.
2. J.E.Kline, Tappi Coating Conf.Proc.,1985, 181.
3. T.Murayama, "Dynamic Mechanical Thermal Analysis of Polymeric Material", Elsevier, Amsterdam, 1978.
4. W.Brown, D.Henley and J.Öhman, Arkiv för Kemi,1964, 22, 189.
5. J.W.Goodwin, in "Solid/Liquid Dispersions", ed.Th.F.Tadros, Academic Press, London, 1987, p.209.
6. A.W.Sisko, Ind.Eng.Chem.,1958, 50, 1789.
7. H.A.Barnes,J.F.Hutton and K.Walters, "An Introduction to Rheology", Elsevier, Amsterdam, 1989, p.19.
8. G.Fadat, G.Engstrom and M.Rigdahl, Rheol.Acta.,1988, 27, 289.
9. G.Engstrom and M.Rigdahl, Nordic Pulp Paper Res.J.,1991, 2, 63.
10. S.E.Sandas and P.J.Salminen, Tappi J., December 1991, 179.
11. P.M.McGenity, J.C.Husband, P.A.C.Gane and M.S.Engley, Tappi Coating Conf.Proc.,1992, 133.
12. J.C.Husband and J.M.Adams, "Interactions in Clay-Based Rotogravure Coating Colours", Nordic Pulp Paper Res.J., in press.
13. L.Jarnstrom, G.Strom and P.Stenius, Tappi J., September 1987, 101.
14. H.C.Black, Anal.Chem.,1951, 23, 12, 1792.
15. R.Pecora, in "Measurement of Suspended Particles by Quasi-Elastic Light Scattering", ed.B.E.Dahneke, Wiley Interscience, New York, 1983, p.3.
16. W.Brown and R.Rymden, Macromolecules,1986, 19, 12, 2942.
17. J.Gregory, in "Solid/Liquid Dispersions", ed. Th.F.Tadros, Academic Press, London, 1987, p.167
18. Th.F.Tadros and A.Zsednai, Colloids Surfaces, 1990, 43, 105

19. D.Heath and Th.F.Tadros, Faraday Discuss.Chem.Soc., 1983, **76**, 203.
20. G.Fadat, I.Pettersson and M.Rigdahl, Nordic Pulp Paper Res.J., 1986, **4**, 30.
21. W.Windle and K.M.Beazley, Tappi J., 1968, **51**, 340.
22. P.A.C.Gane, P.M.McGenity and P.Watters, Tappi J., May 1992, 61.
23. M.Adolfsson, G.Engstrom and M.Rigdahl, Tappi Coating Conf.Proc., 1989, 55.
24. G.Engstrom and M.Rigdahl, Tappi J., May 1987, 91.

Controlled Stress Stimulation of Spray Coating

Mark Power and Steve Smith
CARRI-MED PLC., CARRI-MED HOUSE, GLEBELANDS CENTRE,
VINCENT LANE, DORKING, SURREY RH4 3YX, UK

1 ABSTRACT

Spray coating is a process involving many differing aspects of rheology. The types of conditions that the sprayed material undergoes during spraying are both wide ranging and difficult to simulate meaningfully. In the past, controlled shear rate rheometers were the only means of attempting to monitor the performance of sprayed materials and the limitations of such equipment meant the devising of techniques which were not truly representative of the 'real life' process. However, controlled stress techniques enable the performance of the material to be tested prior to usage and allow the behaviour in the actual process to be predicted much better than previously possible.

2 INTRODUCTION

Many industries use spraying to coat one particular substance onto another. This can be done for a whole host of reasons; spraying paint onto many types of goods is a common example, applying the underbody sealant to cars is another.

Until now the way to test whether or not the sprayed material will stay in-situ on the coated surface has been rather limited because of the lack of suitable rheological equipment some years ago when the spray techniques were first developed. This has resulted in what are now 'old-fashioned' techniques being used in today's environments. Although many would acknowledge that the techniques used are not particularly representative of real-life situations, they would appear repeatable to some extent. However, do repeatable laboratory results necessarily confer process success? The current amount of work being put into investigating the performance of sprayed materials would suggest not. But why is this? Let us look at what happens during the spraying process ......

3 SPRAYING..... WHAT HAPPENS?

A simple spraying system is comprised of a reservoir of the sprayed material, a pipe network to deliver the material from the reservoir to the spray head and the spray head (or nozzle) itself. There is also some means, be it a pump or pressurising of the reservoir, to force the material up to and out of the spray nozzle. See fig. 1 for a simple schematic of a system for spraying the underbody sealant onto the lower surfaces of motor vehicles in a production line.

In simple terms the process is as follows ..... The material originally lies at rest in the reservoir and is forced to move from rest along the pipes to the spray nozzle. Ideally, the viscosity of the material drops under the stress it experiences in the pipework (often reaching an equilibrium viscosity), thus facilitating ease of flow to the nozzle. Finally, it is forced through the nozzle, ejected into the open air to land a very short period later on the surface to be coated. But how does this translate into rheological terminology?

The material, which is initially at rest, has to be forced to yield, i.e. begin to flow. It is then subjected to conditions of constant shear rate (if pumped) or constant stress (if moved by pressure). As the material nears the nozzle it is subjected to many different rheological phenomena - especially if the nozzle is sharply converging - and the material is squeezed through a narrow constriction where it often undergoes quite high shear rates. It is subsequently ejected into the air and onto the surface to be coated. It then stays on the surface, subjected only to the low but constant stress of gravitational effects (i.e. its own weight).

Let us now detail each of these individual rheological contributions to the overall success of the spraying.

Yield Stress of the Material in the Reservoir

Yield stress is a much discussed phenomenon. Whether or not a true yield exists is very much a matter of time-scale. When one studies the rheological behaviour of a sample over many decades of shear rate (typically 10^{-7} to 10^{4} $secs^{-1}$) it becomes clear that the behaviour of most non-Newtonian samples is dictated by two limiting Newtonian plateaux (one at very low shear rates, one at high shear rates), interconnected by a region of shear thinning behaviour where the viscosity drops as a function of the shear rate. However, in many cases this assumes that long time periods are available to measure the low shear limiting viscosity. Clearly, in a spraying process this is not possible as the material must flow within a very short period of applying the driving force, and the existence of an 'apparent' yield value must be acknowledged.

This means that the apparent yield value must be capable of being measured directly and accurately. Accordingly, this means the use of a rheometer that is sensitive enough to measure at very low shear rates. The better the ability of the rheometer to measure at very low shear rates (i.e. to resolve very small displacements of the measuring system), the better the accuracy with which the apparent yield stress will be measured. If the rheometer used is not capable of measuring at very low shear rates (which is usually true of controlled shear rate rheometers and viscometers), then the yield stress has to be obtained by extrapolating data obtained at higher shear rates back

Pressure
A GENERAL SPRAYING APPLICATION
Sealant
Nozzle
Bingham
Shear thinning
Yield stress
STRESS
SHEAR RATE
0
10
20
30
40
50
60
100
200
300

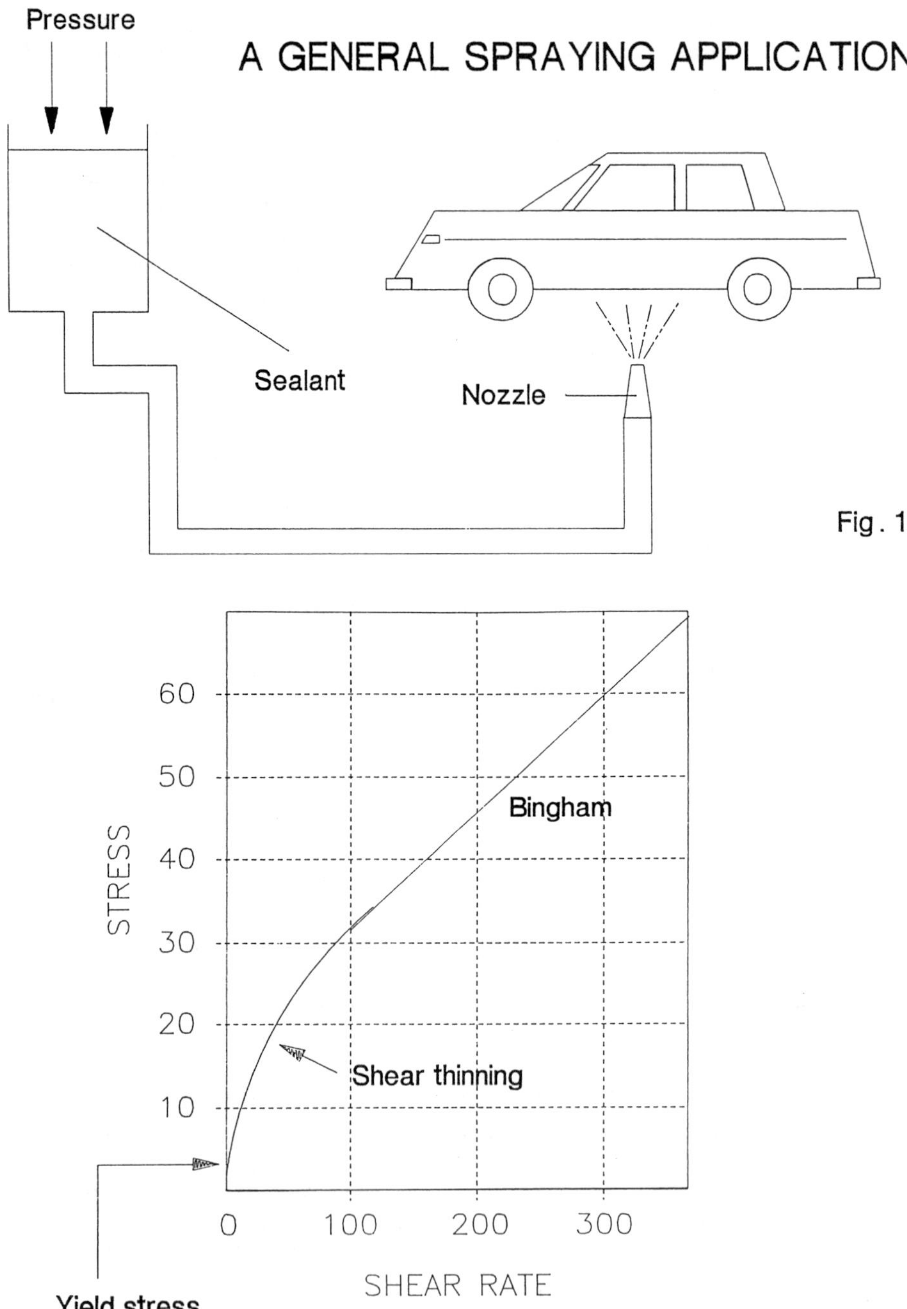

Fig. 1

Fig. 1a

to the zero shear rate axis (on a stress versus shear rate graph) using curve shape prediction techniques. This is inherently an unreliable technique as it depends upon the lowest shear rate that can be measured and, of course, the behaviour of the sample.

Breaking the Structure of the Sprayed Material Sufficiently to Enable it to be Sprayed

Normally, this section of the process is taken very much for granted. The mere fact that the sample has started to move and is capable of being kept moving by the driving force is usually sufficient. However, it is possible to design into the material specific characteristics for the process of breaking down its internal structure sufficiently to pump to the spray nozzle, but to optimise such parameters as power consumption and recovery of the material after spraying. After all, it would be useless to have a material that was extremely easily pumped and sprayed, if its structure was so damaged that it could not recover quickly enough on the coated surface and was lost due to dripping, for instance.

A practical example would be sprayed underbody sealants for cars. Often the spraying is carried out by robots, that do not have the decisive capability to assess when there is too much or too little sealant on the vehicle. Accordingly, the material has to retain enough structure during the pumping and spraying to ensure that it will remain on the vehicle within certain acceptable limits of thickness. It would therefore appear necessary to produce sealants whose viscosity drops relatively quickly when subjected to some form of driving force, but maintain their viscosity above a certain shear rate so that minor changes in the processing conditions do not result in material of a widely varying viscosity deposited on the vehicles. Such behaviour is shown in fig. 1a.

Using either a controlled shear rate or controlled stress rheometer would prove suitable for simulating this stage of the overall process, which normally requires shear rates on the region of 100-1000 $secs^{-1}$.

The process as a whole can be very much a trade-off between several desirable properties such as ease of application, suitable coating thickness, minimising sag after application, good levelling, etc.

The Spray Nozzle

When the material goes through the nozzle it undergoes a very complex rheological experience. Both shear flow and extensional flow occur at the same time, and extensional and visco-elastic effects play a significant part in both the flow through the nozzle itself and the state of the material as it goes through the air to the object surface. The size of the droplets of material as they fly through the air are influenced by these rheological effects and this has a subsequent effect on the state of the material on the object surface. Indeed, much work has been performed in previous years on this subject by Dr Jim Ferguson in the Dept of Pure and Applied Chemistry at Strathclyde University.

To a certain degree, because of the complex nature of this stage, it is possible to neglect what happens during this period if one has the ability to measure how the material performs once it has landed on the surface. Shear rates of 1000-10000 $secs^{-1}$ are usually the norm in flow through a spray nozzle.

When the Material is on the Object Surface

Here the material is subject to completely different conditions. The conditions are purely stress driven, the stress usually being quite low and due to gravitational effects, dependent only upon the density of the material and the thickness of the applied layer. The material will initially have a rippled or mottled surface due to the effect of droplets from the spray, and the degree of unevenness will depend upon the speed and size of the droplets as they hit the surface. Normally, the surface will be expected to level (see fig.2) prior to curing or arriving at the final state, but will then be expected to stay in-situ as long as it takes to arrive at the final state, and often this can be a relatively long period (see fig.3). A significant variation in thickness of the layer can lead to localised areas of high stress differential and the subsequent deformation of the layer, which could lead to flow and eventually to phenomena such as dripping.

Fig.2 shows how the surface is affected by rheological effects, with the levelling of the initially rippled surface dependent upon the surface tension (δ). The stresses prevalent are calculated by using the equation in the diagram.

Fig. 3 shows the tendency of materials on a vertical surface to want to slump down under the effect of gravity. Here the stresses prevalent can be calculated using the material density (ρ) and the thickness of the layer (h). It is also possible to calculate the thickness of the layer at any time t, at any point within the layer after initial application of the material......

For a Newtonian sample :

$$h = (Xn/gt\rho)^{1/2}$$

where .....

h=	thickness at time t
X=	distance from top of slab where thickness is desired
n=	viscosity
g=	9.81 m/s^2
t =	time
ρ=	density

For a Power Law sample :

$$h=(X/t)^{\frac{i}{i+1}}\ (K/\rho g)^{\frac{1}{i+1}}$$

LEVELLING AFTER SPRAY COATING

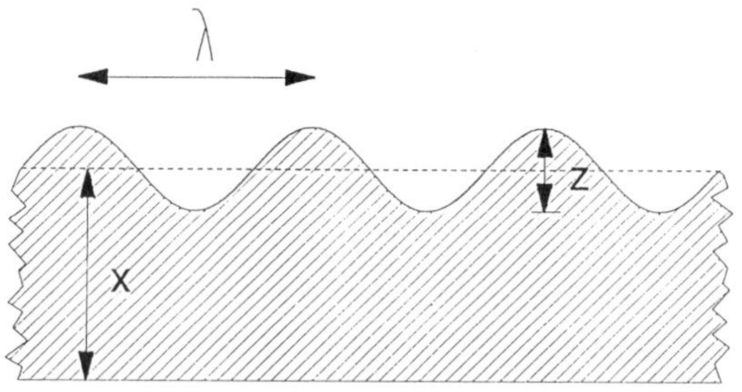

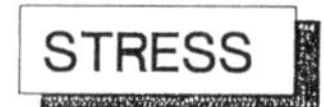

$$\sigma_{MAX} = \frac{248 . ZX\delta}{\lambda}$$

Fig . 2

SAGGING AFTER SPRAY COATING

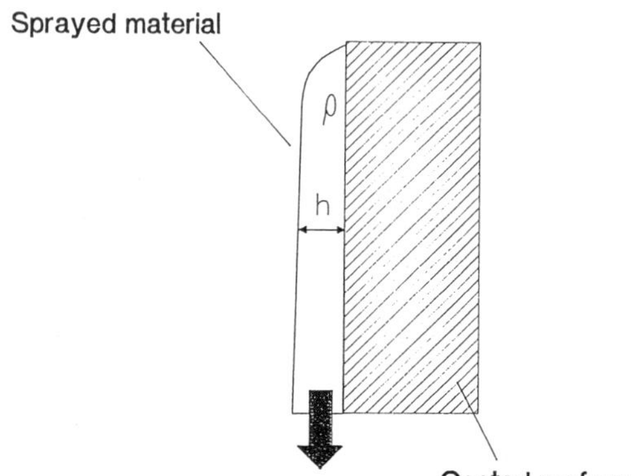

STRESS

$$\sigma_h = h\rho g$$

Fig . 3

where .....

i = power law index
K = viscosity coefficient

Both these examples assume that the material was initially deposited as a slab of uniform thickness.

A controlled stress rheometer can simulate directly the low stresses involved in both the sagging and levelling of coatings, which are often at shear rates in the region of 10^{-4} $secs^{-1}$ or below.

4 WHAT TECHNIQUES ARE CURRENTLY USED TO CHARACTERISE SPRAYING?

At present, the process is simulated by subjecting the material to be sprayed to a short period of increasing shear rate, normally from the minimum the rheometer can apply to a pre-determined maximum (called the UP CURVE), holding that shear rate for a specified period of time (PEAK HOLD CURVE) and then decreasing the shear rate from the maximum back to the minimum shear rate over the same period that it was ramped up (the DOWN CURVE).

When the results are obtained, the significant parameters are the yield values on both the up and down curve and the area between the up and the down curve (see fig.4). The yield value for the up curve is representative of the input required to get the material moving in the reservoir. The area between the two curves is representative of the structure breakdown required to pump the material through the system and the down curve yield point is representative of the state of the material when it is on the surface to be coated, i.e. will it stay there or not?

What are the Problems Associated with this Technique?

Measuring the 'up' yield point is the only way of predicting how the material will respond to the initial force placed upon it. However, as stated previously, it is much better to be able to measure this directly than rely on extrapolated data.

Measuring the loop area between the two curves is of little value when trying to measure the materials under 'real-life' conditions because the material is never subjected to gently reducing stresses during the spraying cycle. Indeed, this technique would seem to suggest that only thixotropic materials are capable of being sprayed, which is clearly not true. What we need to do is to model the flow behaviour during the period of increasing stress, deduce an equation that describes the rate at which the viscosity changes with stress (or shear rate) and use that equation to decide the process parameters to set during spraying.

Using A Flow Curve To Predict Spraying Characteristics

60
50
40
30
20
10
STRESS
Up Curve
Down Curve
0
1
2
3
SHEAR RATE

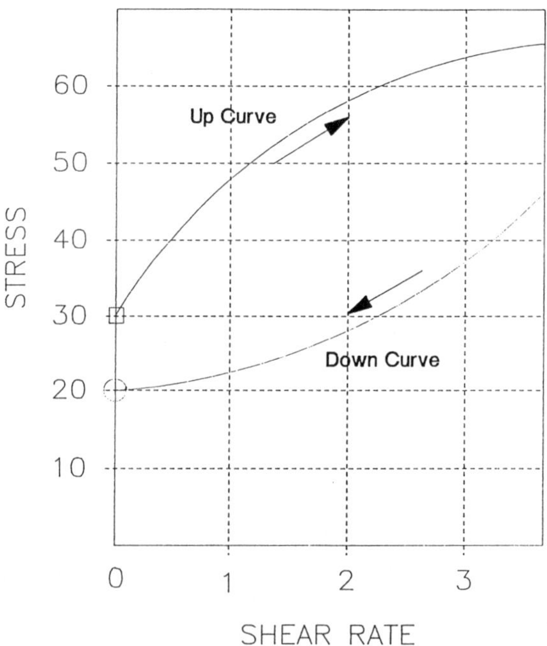

Fig. 4

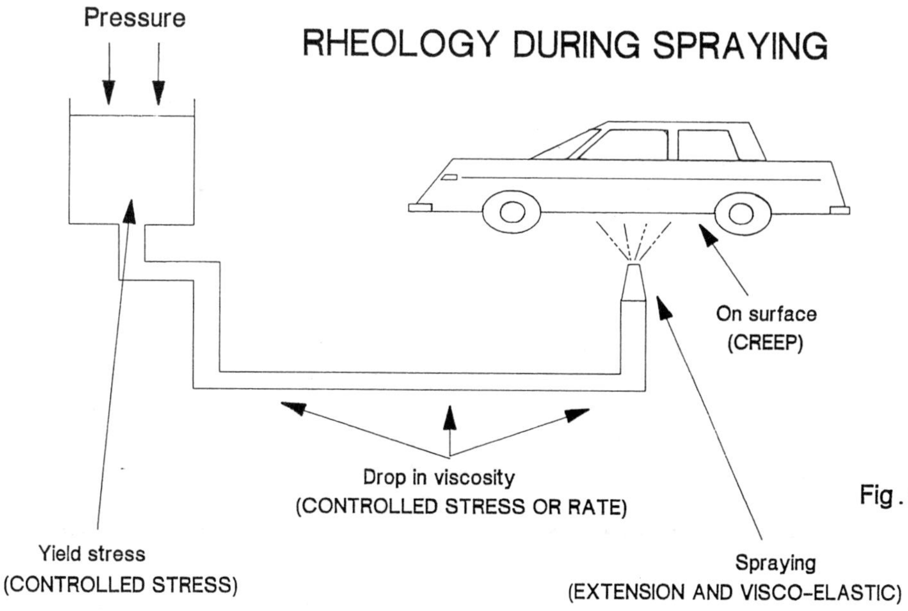

Pressure
RHEOLOGY DURING SPRAYING
On surface
(CREEP)
Drop in viscosity
(CONTROLLED STRESS OR RATE)
Yield stress
(CONTROLLED STRESS)
Spraying
(EXTENSION AND VISCO-ELASTIC)

Fig. 5

Lastly, when the material is ejected from the spray nozzle all forces upon it are removed instantly, with the exception possibly of small stresses due to its flight through the air causing small deformations in the shape of the droplets. The material does not undergo a gentle reduction in stress, as the measurement technique would suggest. It merely hits the surface (causing localised 'splattering' effects) and then settles down under its own weight. The measurement of the down curve yield stress is therefore useless in assessing the ability of the material to recover its structure in-situ on the coated surface. Here we need to use a technique that can instantly reduce the stress to zero and then apply a constant stress equatable to that due to the materials weight.

5 IS IT POSSIBLE TO SIMULATE THE SPRAYING CYCLE?

In a word - Yes. Fig.5 shows the cycle of rheological events that the sprayed material is subjected to during spraying. The Carri-Med Controlled Stress Rheometer is designed to simulate both stress and shear rate driven rheological processes and the computer control of the rheometer allows the linking together of a variety of techniques that simulate exactly what happens in everyday flow and deformation. Spraying as a process is a combination of two completely different rheological techniques; flow and creep.

All the parts of the process leading up to the ejection of the material from the spray nozzle are examples of common flow. As stated previously, the prime driver can be the speed of flow (shear rate driven) or the force (stress driven). The Carri-Med Controlled Stress Rheometer can simulate both of these situations. The part of the spray process involving the material on the surface to be coated is a sample of creep, i.e. the sample is under a single, constant stress and deforms or flows in response to that stress.

How Does a Controlled Stress Rheometer Simulate Spraying?

As in the experiment outlined earlier in this paper, the controlled stress technique starts by increasing the stress (from zero) on the material. However, because a controlled stress rheometer need not be moving to measure it is possible to increase the stress from zero and log the stress point at which the sample first moves. The Carri-Med Controlled Stress Rheometer has a resolution of 10^{-5} radians, and so the point of first movement is easily detected - hence very accurate, direct measurement of the 'apparent' yield stress of the material in the reservoir.

Once this has occurred the rheometer continues to measure the sample by increasing the stress to the pre-determined maximum, again producing a graph of stress versus shear rate. However, the down curve is neglected because, as stated previously, the material is not subjected to this in real-life. Instead, the computer control allows the operator to program the system to turn off the stress at the maximum value and then instantly apply a constant low stress to the sample - thus simulating exactly what happens in the process. The rheometer then logs the deformation of the sample with time and the analysis of the resulting creep curve gives valuable visco-elastic information (taking into account the immediate shear history of the sample) as to how the structure of the material accommodates the constant force upon it - or if it can! Accurate predictions can then be made as to the suitability of the material for use.

The following pages outline the work that I have done for a well known sealant manufacturer who was interested to see what differences there were between two samples of a sprayed underbody sealant used on production cars.

6 A PRACTICAL EXAMPLE OF SPRAY SIMULATION

The plot overleaf shows comparative flow properties of the two samples (SPRAY -1 AND SPRAY -2). As we can see, both show an initial period of shear thinning behaviour when subject to the stress, but then start to behave as Bingham plastics (i.e. have a constant viscosity) after a certain shear rate. As stated previously, this is a good property for some sprayed materials to have as it can mean good spraying characteristics (if the viscosity is low enough), with minimal 'damage' to the sample so that it stands a better chance of staying on the surface.

The analysis of the curves reveals the following:

	SPRAY-1	SPRAY-2
Yield Stress	200.0	168.1
Rate Index (early curve)	0.3321	0.3317
Bingham viscosity (later)	10.03	6.49

From this one would say that SPRAY-2 would be the better sample for spraying applications because it has the lower yield stress, lower rate index and lower Bingham viscosity. This means that it starts to flow easier (less energy input require to get it flowing), it thins down more quickly (easier to get into the pipework, etc. from the reservoir) and it has a lower viscosity at the shear rates likely to be experienced during pumping to the spray nozzle.

But what about the performance of the materials when on the surface of the car:

The plot entitled 'Comparing performance on car' shows that the slumping of SPRAY-2 is considerable compared to that of SPRAY-1, and this is not desirable. It is also contradictory to the flow information where SPRAY-2 would be considered the best. Visco-elastic analysis of the results shows that SPRAY-1 attains a steady state viscosity of 9,585,000 Poise within the 5 minute experiment, compared to only 239,700 Poise for SPRAY-2. It is therefore conclusive that sagging is far more likely with SPRAY-2.

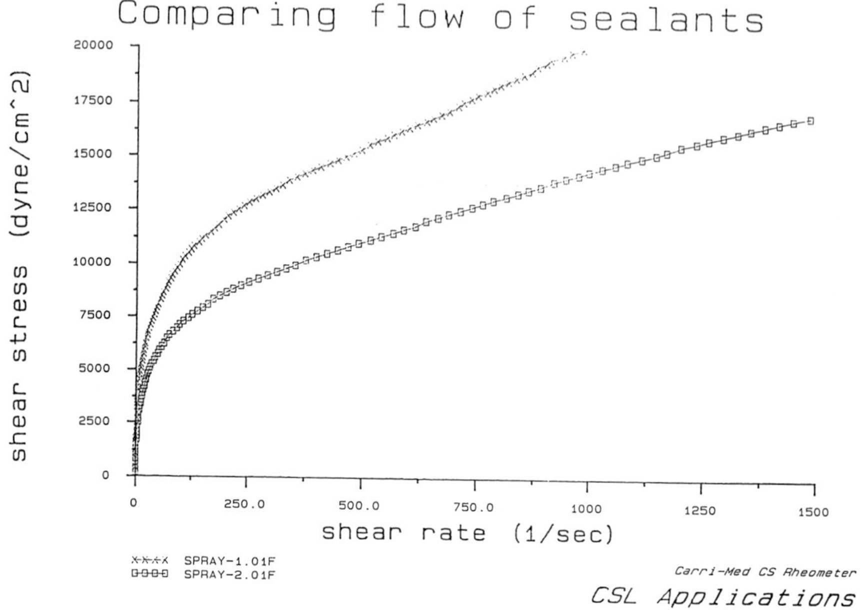
Comparing flow of sealants
shear stress (dyne/cm^2)
20000
17500
15000
12500
10000
7500
5000
2500
0
0
250.0
500.0
750.0
1000
1250
1500
shear rate (1/sec)
SPRAY-1.01F
SPRAY-2.01F
Carri-Med CS Rheometer
CSL Applications

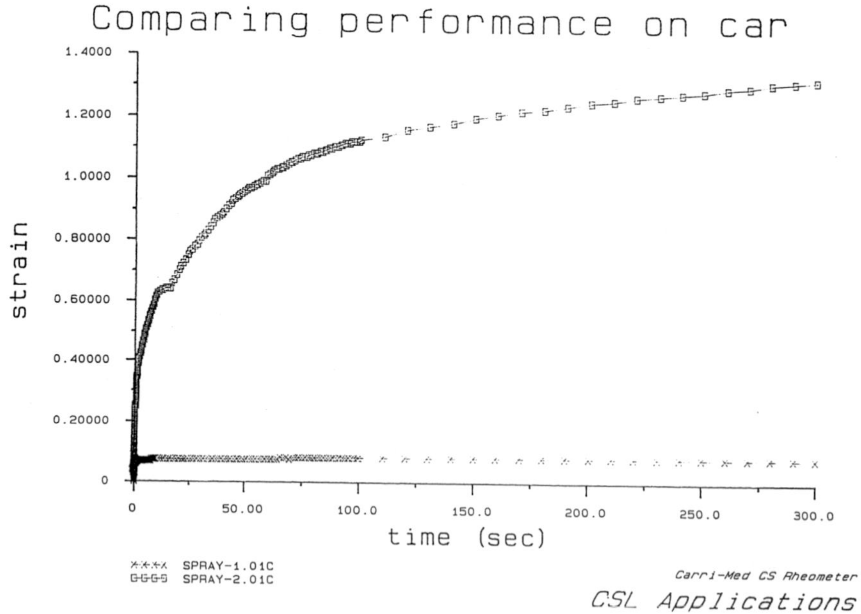
Comparing performance on car
strain
1.4000
1.2000
1.0000
0.80000
0.60000
0.40000
0.20000
0
0
50.00
100.0
150.0
200.0
250.0
300.0
time (sec)
SPRAY-1.01C
SPRAY-2.01C
Carri-Med CS Rheometer
CSL Applications

The overall conclusion of this was that SPRAY-1 would be the best formulation to use for spraying applications. It has slightly worse spraying characteristics, but far better anti-slump and sag properties than SPRAY-2. As usual, a compromise is the order of the day - with a trade-off between one or more of several desirable properties that a material is required to exhibit.

Fluid Mechanics

Classification of Coating Flows

Hadj Benkreira
DEPARTMENT OF CHEMICAL ENGINEERING, UNIVERSITY OF BRADFORD, BRADFORD BD7 1DP, UK

1. INTRODUCTION

The objective of a coating operation is to deposit a flow of liquid on a surface, under prescribed constraints. The range of constraints is very wide:

(i) the surface may be a solid or another liquid (multi layer coating),flat or shaped, smooth or rough, stationary or moving, impermeable or porous, etc.

(ii) the liquid may be a paint, an adhesive, an ink, a solvent or water based suspension, a polymeric formulation, etc.

(iii) the thickness of film required may be a few microns or of the order of millimeters,

(iv) the speed of production (substrate speed) may be of the order meters per minute to meters per second,

(v) the standard required for the quality of the film (uniformity, wallpaper).

Clearly a variety of coating methods are available depending on application and these inevitably evolved in isolation with each other, leading to a fragmentation of the technology and little exchange in this important and rapidly growing industry. A classification of coating methods is thus a logical starting point to establish a framework (or analysing coating flows). Two categories can immediately be discerned:

(i) continuous coating flows, where a flow is applied on a moving surface,

(ii) "batch" coating flows, where a flow is applied on a discrete surface, an object, a car or a toy for example.

This paper deals with continuous coating flows which in themselves are varied and have received little ordering attention. A further relaxation of the constraints is to assume the moving surface to have a uniform geometry, ie a flat surface, but to account for porosity, roughness and surface energies which distinguish the coatability of surfaces.

The ultimate aim of this, now well defined but general problem is to:

- deposit a liquid of given properties in the guise of a film of prescribed thickness and uniformity at a prescribed speed, on a flat surface of given properties.

Such a task can be carried in one or a combination of the following broad ways:

i) plunging the surface in a bath of the liquid available and withdrawing it at the prescribed speed and hoping the constraints on film thickness and uniformity are met, ie free drawal coating

(ii) metering an excess amount of liquid on the surface in a flow geometry to meet these constraints, ie metering or constrained coating

(iii) delivering via a flow geometry the exact amount of liquid required for the surface, ie die or exact coating

(iv) brushing, printing or gravure coating onto a substrate the exact amount of liquid required.

This intuitive methodology forms the basis of coating flows classification which is now examined further.

2. FREE WITHDRAWAL COATING

A general operating scheme of this kind is shown in Figure 2.1 where the moving surface is allowed to enter and leave the liquid bath at different angles. Apart from the prediction of film thickness and uniformity in relation to operating variables the onset and extent of air entrainment must also be of interest. An immediate shortcoming of this operation is that both sides of the substrate are coated. This may not be feasible and variations of such an operation are shown in Figure 2.2.

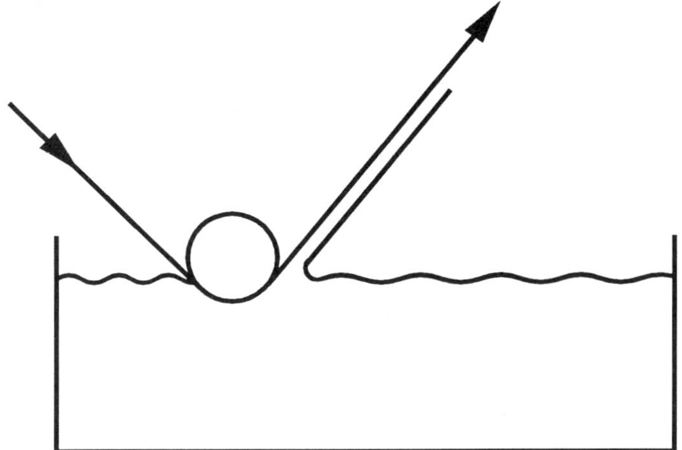

Figure 2.1 Withdrawal Coating

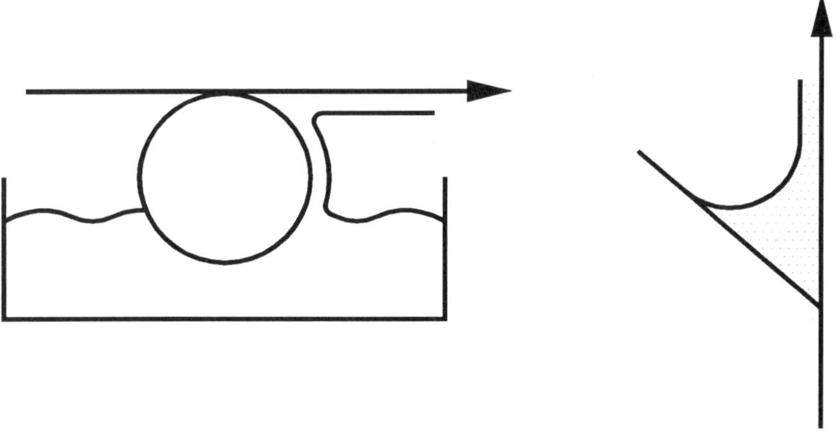

Figure 2.2 Variation Of Standard Withdrawal Coating

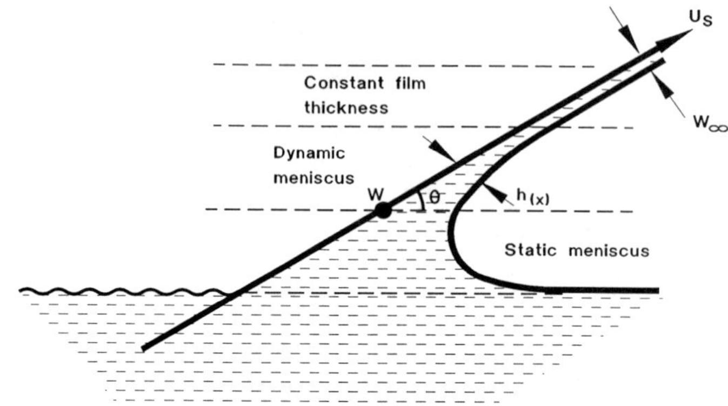

Figure 2.3: Flow Analysis Geometry of Free (Withdrawal) Coating

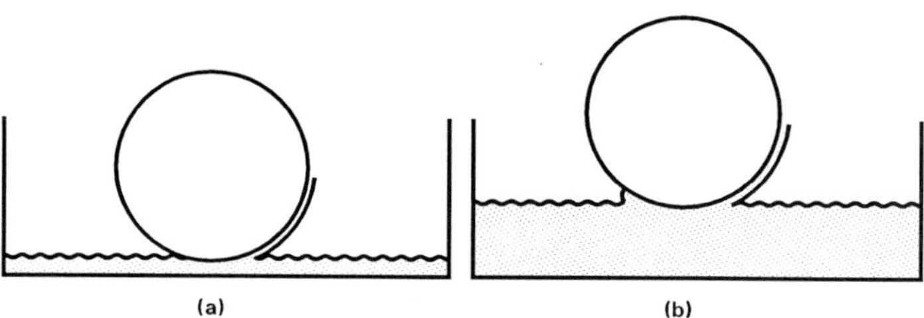

Figure 2.4 & 2.5 (a) Effect of immersion depth on film thickness
(b) Meniscus coating

In all cases the analysis of flow leads to considering the general situation depicted by Figure 2.3 where the film thickness on the substrate is seen to vary with position on the substrate as well as operating variables (fluid properties, substrate speed, angle)

One important controlling variable often ignored is the size of the pool of liquid available. If small enough, it must produce very thin films as shown in Figure 2.4. We are now constraining flow by not presenting a sufficiently large pool.

We can push this constraint further by allowing the moving surface just to touch the surface of the liquid in what is termed meniscus coating, Figure 2.5. Variations of this type of flow are also shown in Figure 2.5.

The analysis of such flow situations must take into account the effect of the surface tension of the liquid, the driving force behind the flow. Clearly no further reduction in film thickness can be achieved without metering or constraining the flow.

3. METERING OR CONSTRAINED FLOW COATING

Here a boundary is put in place to reduce the extent of the free surface. The boundary may be a solid knife or an air-jet, both capable of adopting various shapes as shown in Figure 3.1.

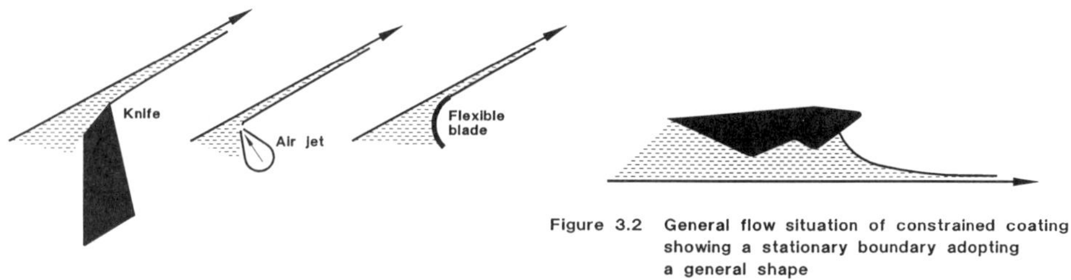

Figure 3.2 General flow situation of constrained coating showing a stationary boundary adopting a general shape

Figure 3.1 Metering or constrained flow coating

In their general form, these flow situations can be represented as in Figure 3.2 where the shape (geometry) of the applied boundary has a great bearing on the thickness and uniformity of the emerging film.

Clearly for a given substrate speed and fluid we can reduce thickness in these situations by:

reducing the gap between the boundary and the moving substrate

but also by :

allowing the boundary to move in the opposite direction to the substrate.

We have now arrived at reverse roll coating (Figure 3.3) where the moving boundary in the shape of a roller is convenient. Note that for convenience also, a roller drives the substrate.

We may also drive the moving boundary in the same direction as the substrate but in a way so that the moving boundary "pinch" some of the liquid which would otherwise end up on the substrate. We have here forward roll coating as depicted in Figure 3.4 where the applied boundary and the direction of the substrate diverge.

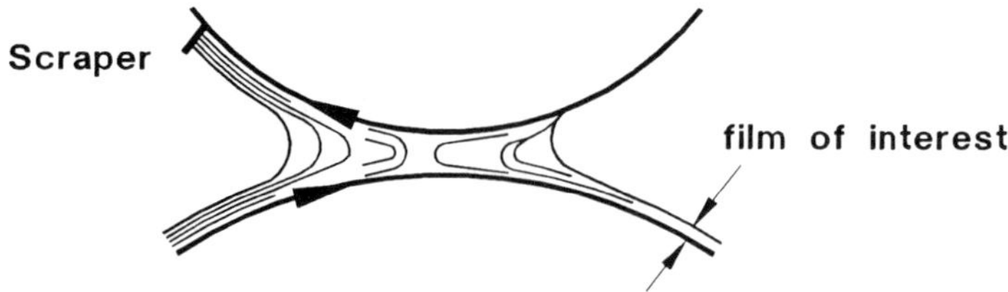

Figure 3.3 A reverse roll coating unit operation. Note the upstream and downstream meniscus may adopt a variety of shapes and positions

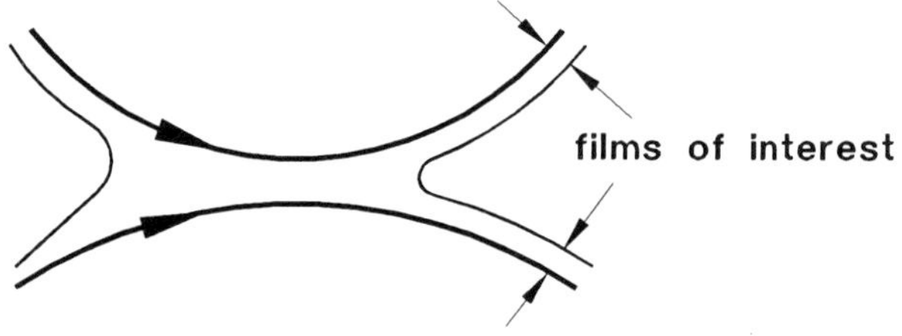

Figure 3.4 A forward roll coating unit operation. Again upstream and downstream meniscus positions and shapes vary with operating conditions

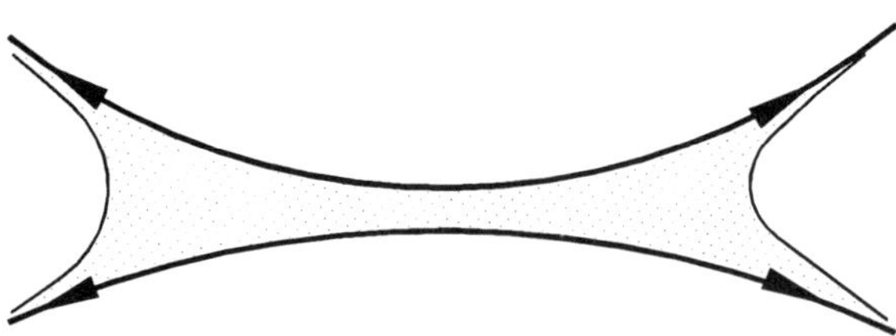

Figure 3.5 General roll coating flow system. The position and shape of the upstream and downstream meniscus are part of the solution to the problem.

Both forward and reverse roll coating can be represented for analysis by the general picture shown in Figure 3.5.

Figures 3.2 and 3.5 both show that the driving flow forces are pressure and drag forces. No other means are available to us to reduce the thickness further except that we may exert a vacuum upstream of the boundary or across it (if it were porous) to reduce the pressure force contributions, (Figure 3.6).

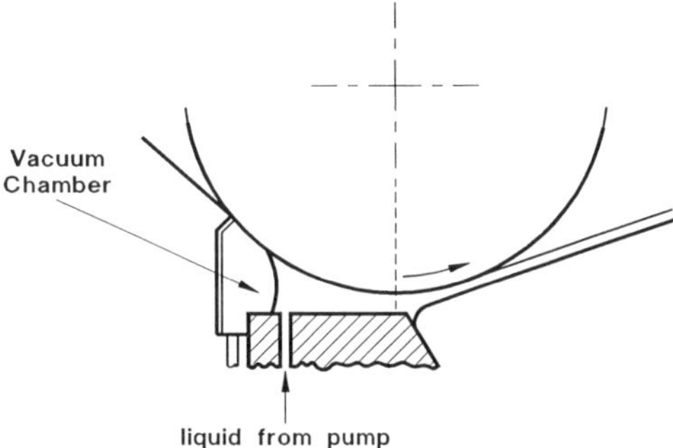

Figure 3.6 : Arrangement for Operation under slight Vacuum

4. EXACT OR DIE COATING

Here we present the substrate with the required amount of liquid. We must therefore adjust the operating variables to this requirement. In slot coating, Figure 4.1, the distribution of the liquid across the width of the die becomes important as the liquid is "free" once out of the die.

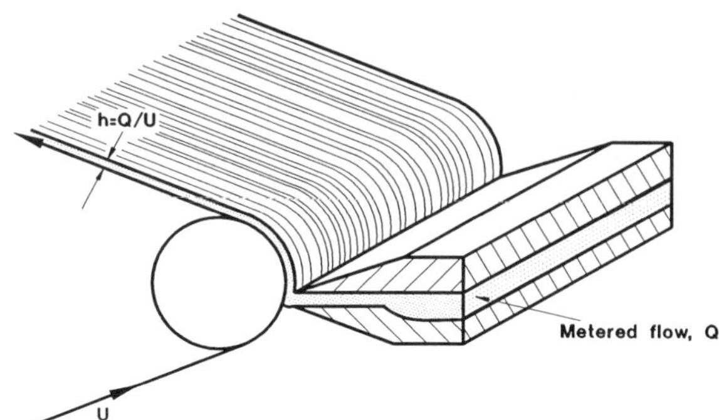

Figure 4.1 Slot Coating required precise die design for uniform flow distribution

In curtain coating, Figure 4.2, a uniform distribution across the width of the substrate is again a difficult requirement to meet as is the stability of the ensuing film. Air entrainment, breakage of the flow are typical limitations.

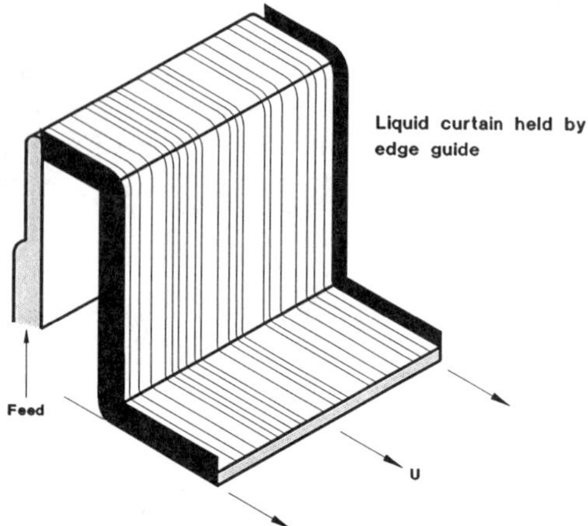

Figure 4.2 Curtain coating require precise die design for uniform flow distribution

On more sophisticated die design, referred to as slot-die coating Figure 4.3, this distribution and stability "window" can be enlarged by manipulating the intricate relationship between the geometry of the die and operating pressures. Here, the liquid emerging from the die is constrained by the die geometry prior to depositing into the substrate.

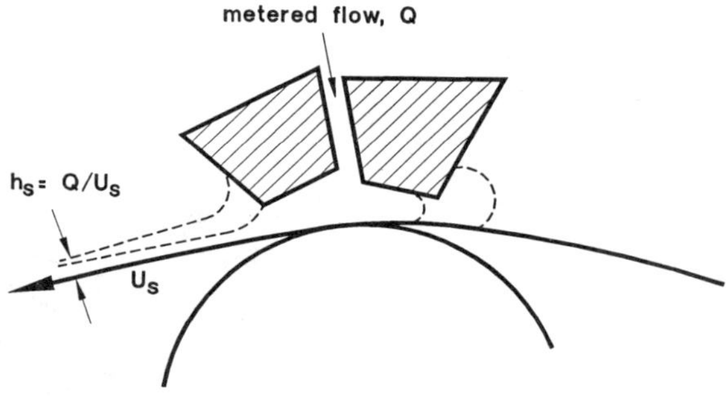

Figure 4.3 Die Coating showing varying bead size

5 GRAVURE COATING

Clearly, using the above arrangement (slot, curtain, die, etc...), there will be a limit to the thickness of the film formed. To reduce the thickness further, we may use gravures or cells to trap the liquid and then transfer it to a substrate, Figure 5.1.

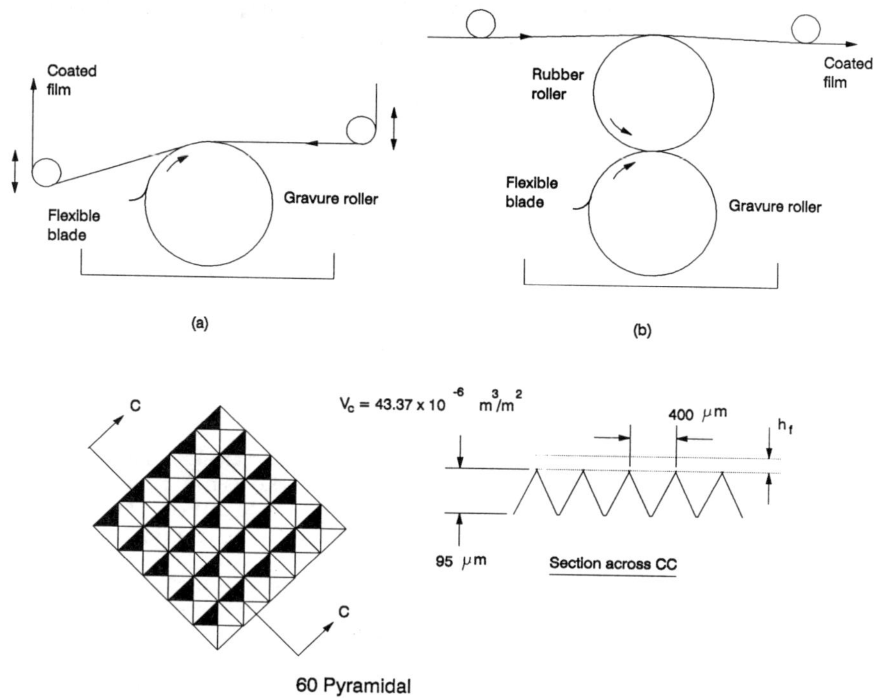

Figure 5.1 Direct (a) and Indirect (b) Gravure Coating and Typical Cell Geometry

This "printing" coating is also produced by brushing (painting) where the liquid is dragged over the substrate by tangential stresses from in between the hair of the brush. Clearly the size and shape of the cells dictate the thickness of the film produced which when forming (direct gravure coating/indirect gravure coating) is prone to instability and air entrainment like all other coating flows.

6. CONCLUSION

Despite their apparent diversity, coating flows can be ordered and classified for ease of analysis. The classification presented enables the application of known results, observation from a particular coating operation, to a number of other operations which may appear to be "different" from one another.

Transfer of Fluids To and From Rollers — Film Thickness in Reverse Roll Coating

D.C-H. Cheng

CONSULTANT IN APPLIED RHEOLOGY, 191 ICKNIELD WAY, LETCHWORTH, HERTFORDSHIRE SG6 4TT, UK

SUMMARY

New experimental results on film thickness in reverse rolling using Newtonian and non-Newtonian fluids are presented and compared with the literature. Theory predicts that the reduced fluid flux through the nip depends on speed ratio only. It is shown that this is not generally valid and the reduced flux depends also on the reduced film thickness on the reverse-rolling roller which returns fluid back into the nip region. The results are also relatively insensitive to non-Newtonian properties. It is pointed out that fluid inertia has an important role in this behaviour. The implication for further development of the theory is discussed. An empirical equation is obtained that can be used for practical application to predict film thickness in industry.

1 INTRODUCTION

The interest and reasons for investigating roll coating processes in general and reverse rolling in particular have been discussed by many authors[1-4]. There is no need to repeat them here. This paper describes work that has since been done at Warren Spring Laboratory (WSL), under the aegis of the Co-operative Programme on Thick Liquids and Pastes: Processing and Flow (TLP). The topic of film thickness in reverse rolling is considered here theoretically, but the work is mainly experimental, involving Newtonian and a number of non-Newtonian fluids. The results are compared with the literature and an empirical equation is also given which can be used for the prediction of film thickness for practical application.

2 THEORY

Previous Results

The theoretical modelling of the film thickness in reverse rolling of Newtonian liquids has been developed by several authors[1,3,4] by solving the one-dimensional Reynolds lubrication equation. All predicted that the film thickness, in the notation of this paper (Fig. 1), is given by:

$$X = Q^{=}(1-Rs) \tag{1}$$

where

$$X = 2(\lambda_2 - Rs\lambda_1) \tag{2}$$

$$\lambda_1 = \frac{h_1}{H_o} \qquad \lambda_2 = \frac{h_2}{H_o} \qquad Rs = \frac{U_1}{U_2} \tag{3}$$

Different assumptions were made by the different authors, leading to different values for $Q^{=}$. Jurewicz[1] obtained 1.0; Greener and Middleman[3], 1.226; and Benkreira et al.[4], 1.24 (+/-2.7%). Using other assumptions from the literature, Greener and Middleman gave another value, 1.30. As a matter of detail, the derivations of Jurewicz and Benkreira et al. were for the case λ_1=0.

These solutions were described by Jurewicz as the "fully developed flow". He recognised that the flow in the roller nip is a very rapid one and there may not be time for the liquid to attain the fully developed velocity due to liquid inertia. Allowing for the liquid acceleration, he found (for the case λ_1=0):

$$X = 2\frac{h_m}{H_o}\left[\frac{1}{2}(1-Rs) + \frac{4}{\pi^2}(1+Rs)\exp\left(-\frac{\pi^2\nu t}{h_m^2}\right)\right] \tag{4}$$

where h_m is the roller surface separation where the pressure is maximum, t time and ν kinematic viscosity. It was not possible to evaluate h_m and t as they are complex functions of roller speeds and kinematic viscosity. Jurewicz argued qualitatively that at low Rs inertia effect is small and the flow is fully developed. The effect is felt at high Rs. The retardation due to inertia means that less liquid is dragged back into the nip by roller 1 and hence more remains on roller 2, giving a larger λ_2 and X.

This Work

It is seen that eqs (1) and (2) are easily obtained by a mass balance:

$$U_2 h_2 = U_1 h_1 + Q \tag{5}$$

where Q is the flux per unit roller width through the nip. Eqs (1) and (2) are the dimensionless form, with

$$Q^{=} = \frac{Q}{H_o U^{=}} \qquad U^{=} = \frac{1}{2}(U_2 - U_1) \tag{6}$$

It now remains to determine Q^-.

It is also easily seen that the theory for forward rolling can easily be applied to the reverse rolling case simply by allowing the speed of roller 1 to be negative. In the notation of this paper, U_1 is the magnitude of the speed of roller 1; hence the negative sign in the expression of U^- in eq. (6). As part of the TLP work, Cheng[5] considered the consequences of making different assumptions systematically. It was shown that in the case of fully flooded up-stream, the value of Q^- depends on the position of the down-stream cavity. It is however trapped between 1.226 and 1.333. In this paper, the average of 1.28 is adopted for comparison with the experimental results.

3 EXPERIMENTS

Previous Work

In the experiments by Benkreira et al.[2], roller 2 was used to pick-up the fluid from a reservoir and transferred to roller 1. The up-stream side of the nip would be described as flooded. A scraper was used on roller 1 and h_1 and λ_1 were taken to be zero. The film thickness h_2 was measured. The speed ratio Rs covered was 0.125 to 1.0. They found that the experimental Q^- = 1.26, with a standard deviation of 0.5%, for Newtonian liquids. For a non-Newtonian fluid, a 1% CMC solution in water, Q^- was higher at 1.38.

In the experiments by Greener and Middleman[3] both rollers were immersed in the fluid reservoir on the up-stream side, and can be described as fully flooded. Both h_1 and h_2 were measured. The speed ratio covered was Rs from 0.25 to 1.75. Newtonian liquids and some non-Newtonian fluids (0.25-0.50% aqueous polyacrylamide) were used. There did not appear to be any systematic difference between the Newtonian and non-Newtonian results. They found that for Rs up to about unity, the results agreed well with their theoretical Q^- = 1.226. In the case of Rs=1, all data lay within +/- 10%. However, as Rs became greater than unity, the experimental X became less than theory. In the case of Rs = 1.75, it was about -2, instead of the theoretical -0.92.For later reference, the authors did not report the values of λ_1 involved.

In the experimental arrangement of Jurewicz[1], liquid was picked up onto roller 2 and transferred to roller 1, which was scraped. The film thickness h_1 was taken to be zero and h_2 measured. Speed ratios between 0.2 and 1.6 and the Newtonian liquids were used. Summarised results were presented in the paper, showing that h_2, ie X, was close to the theoretical 1.0 for small Rs, but as Rs was increased at constant U_2, X decreased somewhat and then increased. The (λ_2,Rs) curves for different U_2 appeared to cross over at a common point. The authors reported that the individual curves and the position of the cross-over point depended on liquid viscosity.

This Work

Apparatus. The roller transfer rig (Fig. 2) was designed and constructed at WSL. It consisted of an interchangeable upper roller 1 (A1), diameter 50 to 215 mm, which can be run in both forward and reverse directions. In the reverse rolling experiments a scraper (not shown) was used to control the amount of fluid returned to the nip. The lower roller 2 (A2) was fixed at 215 mm diameter. This was used to pick up fluids from a reservoir (T) and transferred onto the upper roller. A doctoring blade (K) was used to control the pick-up. The rollers were independently driven (G1, G2 not shown) and the speeds could be varied up to 7 m/s with the 215 mm rollers. The nip gap could be varied (E) from 50 to 1000 μm; it was set with the help of feeler gauges.

Visual or photographic observations were made on the fluid films, particularly when investigating film stability, the on-set of ribbing and the nature of uneven films. The film thickness was measured using calibrated capacitance probes (R).

In the reverse rolling experiments, the gap width was first set and then, with one roller speed held constant, the other was varied. Three conditions of doctoring were used: either with the blade very close to the roll surface so that the nip was nearly starved, or with it fully withdrawn to flood the nip; in the third intermediate condition, it was adjusted (if necessary during a run as the speed was changed) so that the width of the fluid run-back, as it returned to the reservoir, was half the width of the roller. Three conditions of scraping were used: a Neoprene scraper was used to give "hard" scraping in an attempt to reduce h_1 to zero (two versions were used with the rubber sheet flexed in the direction of or against the rotation of the roll surface), or a metal plate scraper was used set at 100 μm from the roll surface, and the third condition was without scraping. It was found that the constant scraping condition did not give constant h_1. In all cases, h_1 was measured as well as h_2.

Materials. Six fluids were used. Their viscosities were measured using Contraves Rheomat 15 or 30 viscometers (covering approximately 3 to 1000 s^{-1}) and a Davenport Extrusion (tube) Rheometer (2.5k to 250k s^{-1}), and shown on Fig. 3. Different batches of some of the fluids were used. Others showed ageing or degradation with re-use. The temperature also varied in the experiments, between 15 to 22 C. The measured flow curves therefore showed variations and Fig. 3 is given as indicative only.

Two lubricating mineral oils, supplied by Castrol Ltd were used. The low viscosity Hyspin 80 oil was taken to be Newtonian, but the high shear rate results showed shear

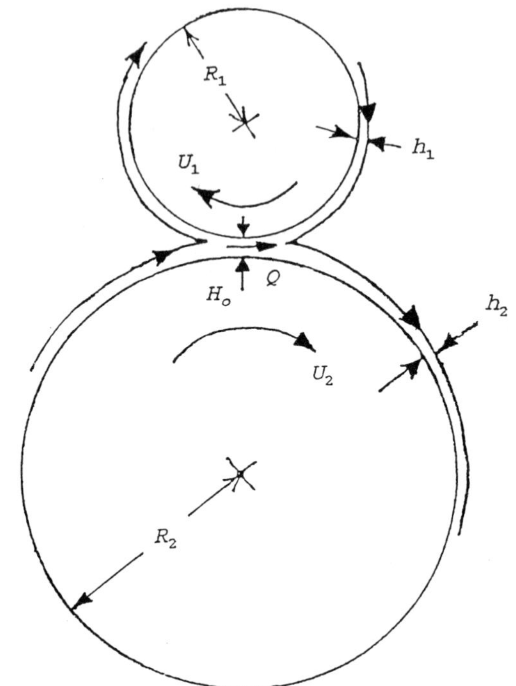

Figure 1. Notation for Reverse Rolling

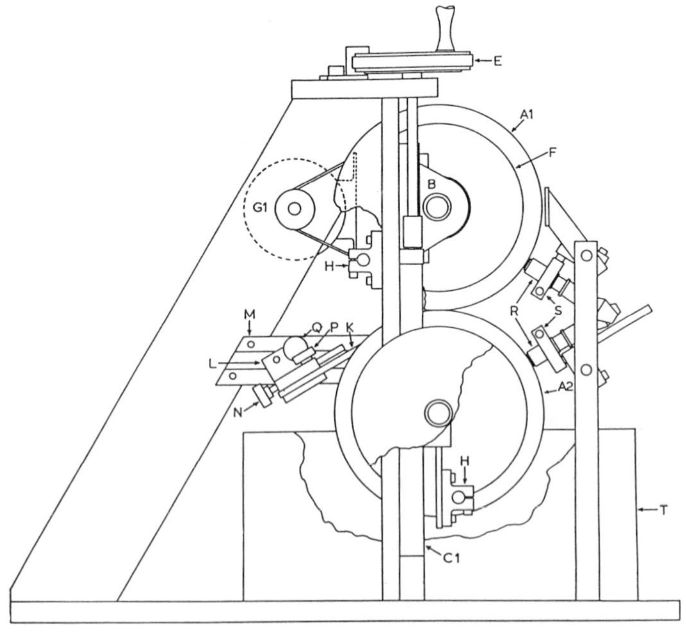

Figure 2. The Experimental Roller Rig - Side View

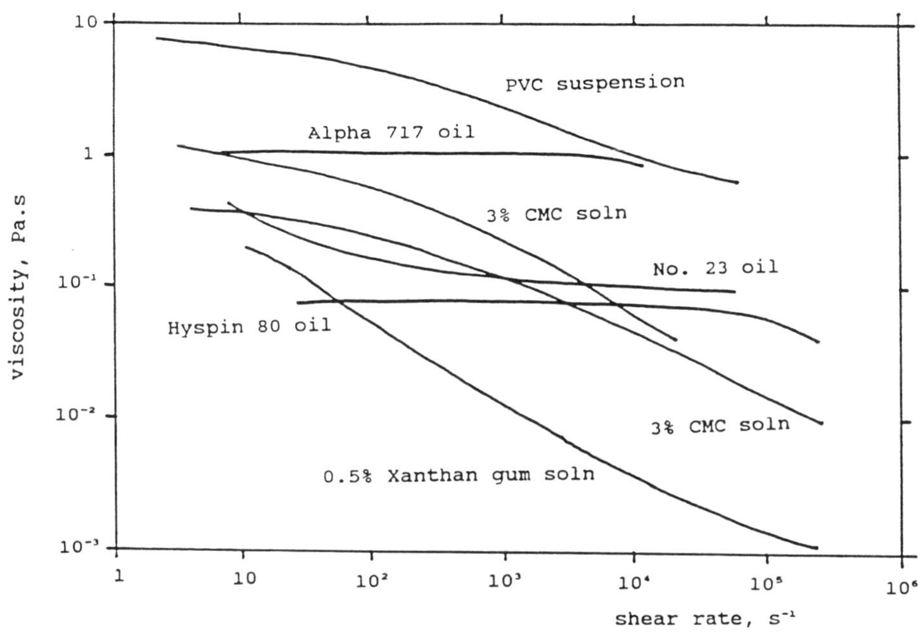

Figure 3. Flow Curves for the Experimental Fluids

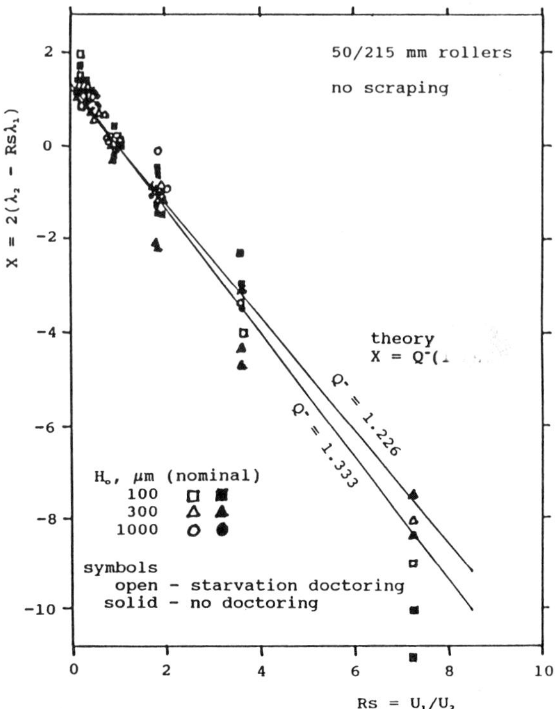

Figure 4. Film Thickness Results for Hyspin 80 Oil - I

thinning behaviour. The high viscosity Alpha 717 oil also showed an indication of shear thinning at high shear rates.

The No. 23 oil was a "non-creep" lubricant, composed of 98.2% solvent refined mineral oil, with polymer additives, and supplied by Burmah-Castrol (UK) Ltd for use as a conveyor lubricant in the textile industry. It became visibly lumpy with time and use, and showed separation, which were signs of ageing and degradation.

The CMC solutions were 3% in deionised water made from powder supplied by BDH Chemicals Ltd. The flow curves varied with preparation and ageing. The range on Fig. 3 shows the variation involved.

The 0.5% Xanthan Gum solution was prepared from powder supplied by Tate and Lyle Ltd and contained 0.1% of Nipacombin preservative.

The PVC resin suspension was an aqueous mixture containing gypsum filler, supplied by Evode Ltd and designated 168. It was used as received.

Results. Typical results are shown for the Hyspin 80 oil. Figure 4 shows that there was reasonable agreement with theory in the absence of scraping, but with the scraper set at 100 μm some results deviated from theory (Fig. 5). With hard scraping, most of the results were very different from theory (Fig. 6).

On replotting the results against λ_1, it is seen that X depends on both Rs and λ_1. At low λ_1, X was more or less constant at X_o =1.28 for all Rs. As λ_1 became large, X tended to the theoretical value X_∞ =1.28(1-Rs). An empirical equation was developed which appeared to fit the data (Fig. 7):

$$\psi = \frac{1}{1+(5\lambda_1)^2} \qquad (7)$$

where

$$\psi = \frac{X - X_\infty}{X_o - X_\infty} \qquad (8)$$

$$X = 1.28 \qquad X_\infty = 1.28(1-Rs) \qquad (9)$$

Similar results were obtained for the other fluids. The extent of fit using eq.(7) is illustrated in Figs 8 and 9 in which data for all fluids for a given roller pair were plotted irrespective of Rs. The scatter appears to be large because the use of ψ magnified the spread. It is not really excessive by the following argument.

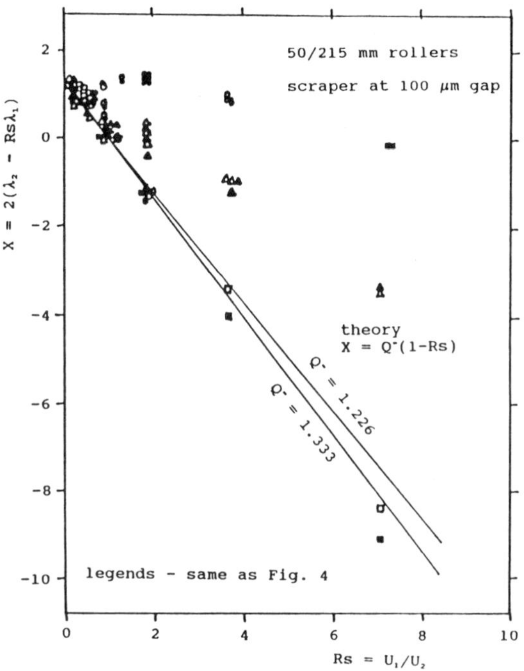

Figure 5. Film Thickness Results for Hyspin 80 Oil - II

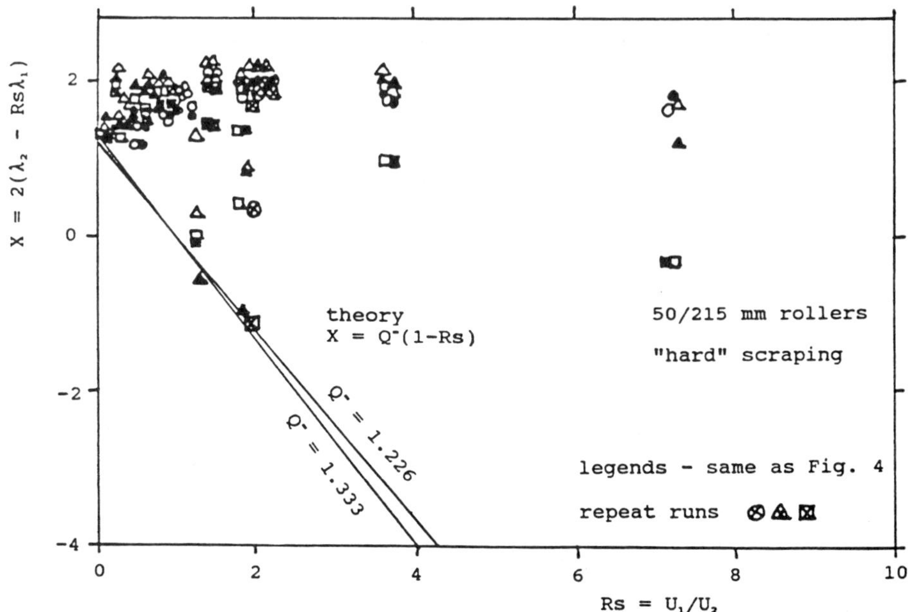

Figure 6. Film Thickness Results for Hyspin 80 Oil - III

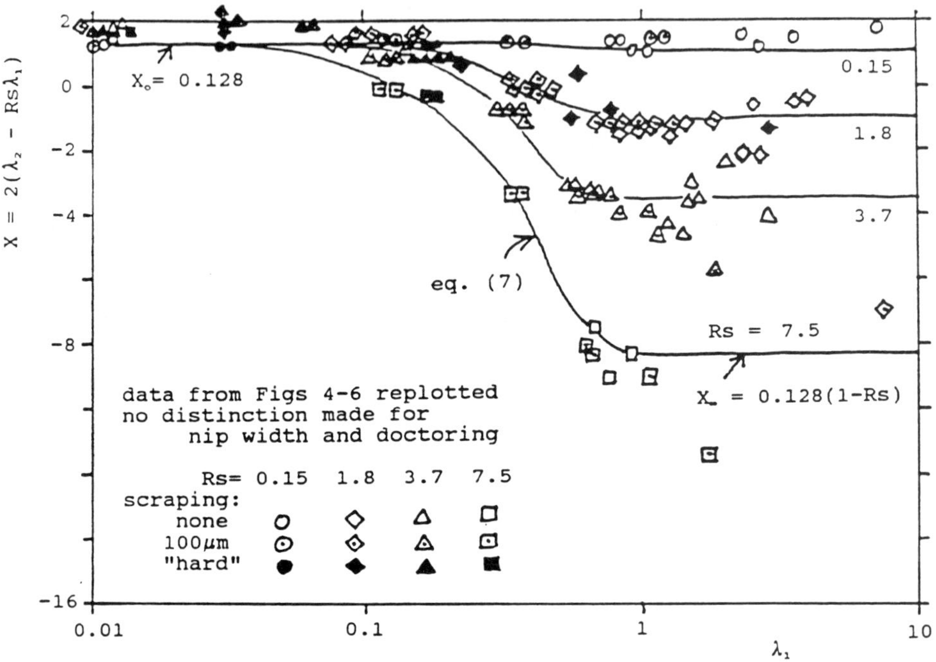

Figure 7. Film Thickness Results for Hyspin 80 Oil - IV

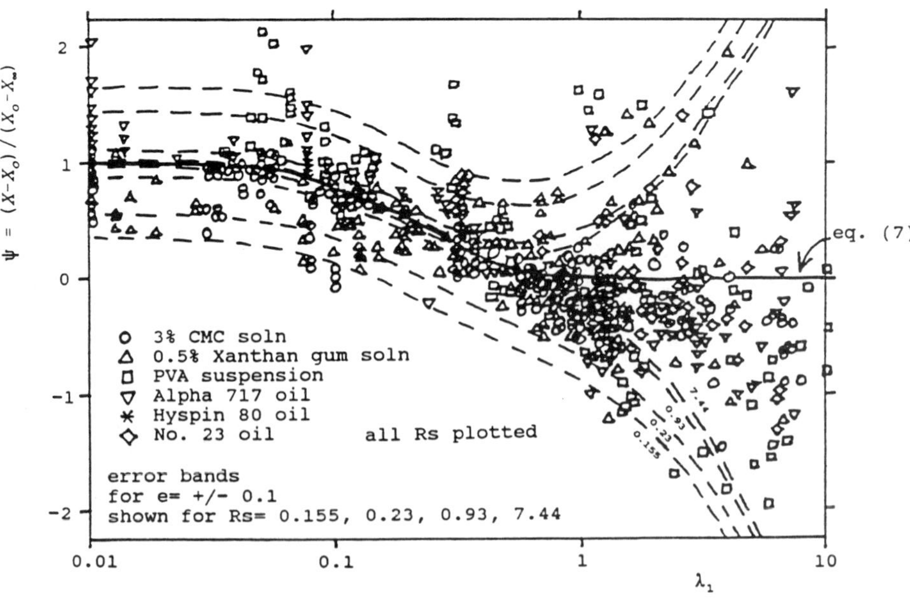

Figure 8. Plot of ψ vs λ_1 for 190/215 mm Rollers

If the experimental quantities were subject to errors according to:

$$\lambda_1 = \lambda_1^{=}(1-e_1)$$
$$\lambda_2 = \lambda_2^{=}(1+e_2) \qquad (10)$$
$$Rs = Rs^{=}(1-e_3)$$

in which $\lambda_1^{=}, \lambda_2^{=}, Rs^{=}$ are the error-free values, then:

$$\psi = \psi^{=}(1+e_2+e_3) + \frac{\lambda_1^{=}}{0.64}(e_1+e_2+e_3) + \frac{e_2}{Rs^{=}} - (e_2+e_3) \qquad (11)$$

Assuming that $e_1=e_2=e_3=e$,

$$\psi = \psi^{=}(1+2e) + \left(\frac{3\lambda_1^{=}}{0.64} + \frac{1}{Rs^{=}} - 2\right)e \qquad (12)$$

On Figs 8 and 9, error bands for e = +/-10% have been drawn. It can be seen that the scatter was more or less evenly spread about eq. (7) and can be attributed to experimental scatter of +/- 10% or 20%, which is not excessive by any means.

All the experimental results are replotted as X vs λ_1 for each Rs irrespective of roller pair and material in Figs 10-13. Equation (7) is also drawn together with the error bands for +/- 10% and 20%. These are given by:

$$X = X^{=}(1+e_2) + 2Rs^{=}\lambda_1^{=}(e_1+e_2+e_3)$$
$$= X^{=}(1+e) + 6Rs^{=}\lambda_1^{=}e \qquad (13)$$

While the low Rs results were fitted well by eq. (7) within the scatter, the high Rs results showed a systematic deviation from the equation. For these, Figs 12 and 13 show that as λ_1 exceeded unity, X dipped and became progressively less than the theoretical X_∞.

4 DISCUSSION

Comparison with Previous Experimental Work

The experiments of this work brought out the dependence of λ_2, ie X, on λ_1, in addition to Rs, and it is rather complicated. It is seen that the fully-developed flow theory

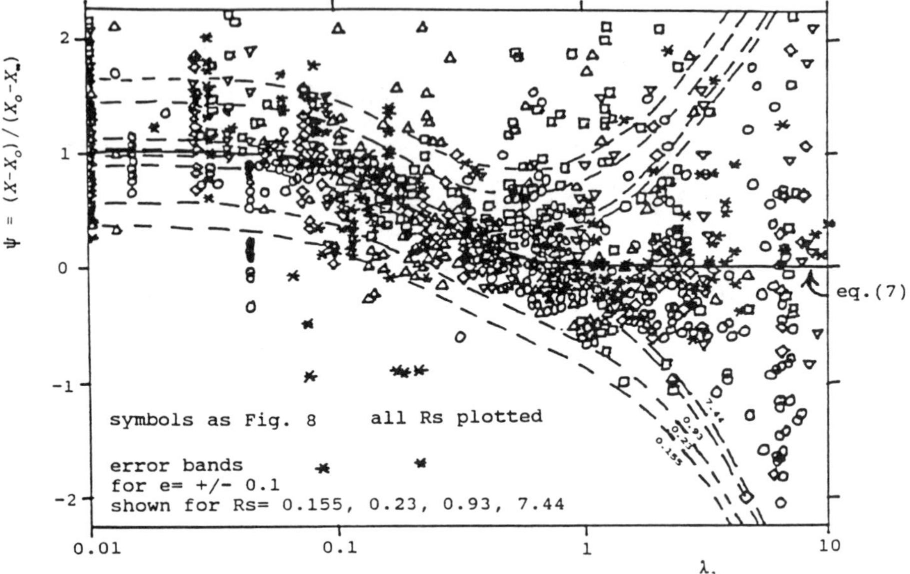

Figure 9. Plot of ψ vs λ_1 for 50/215 mm Rollers

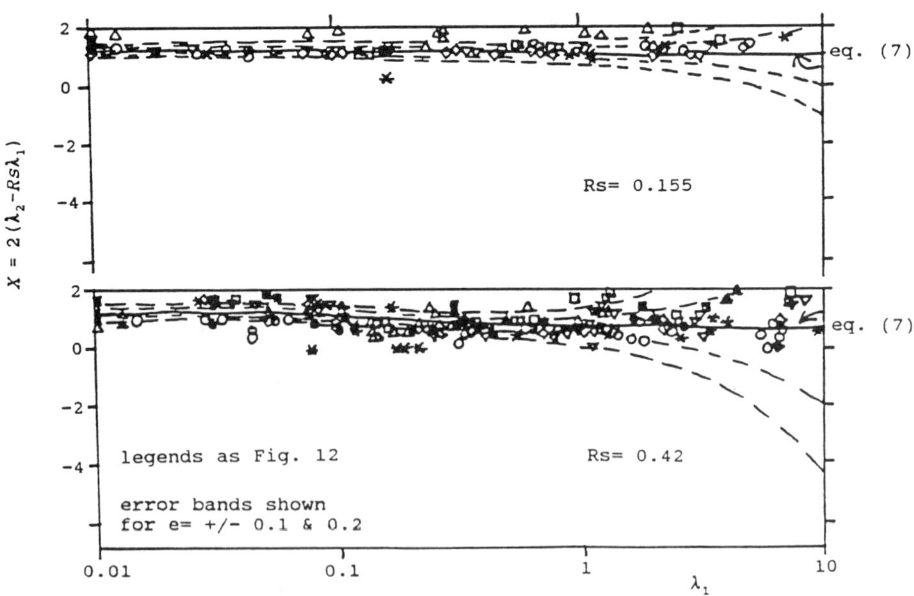

Figure 10. Plot of X vs λ_1 for Rs = 0.155 and 0.42

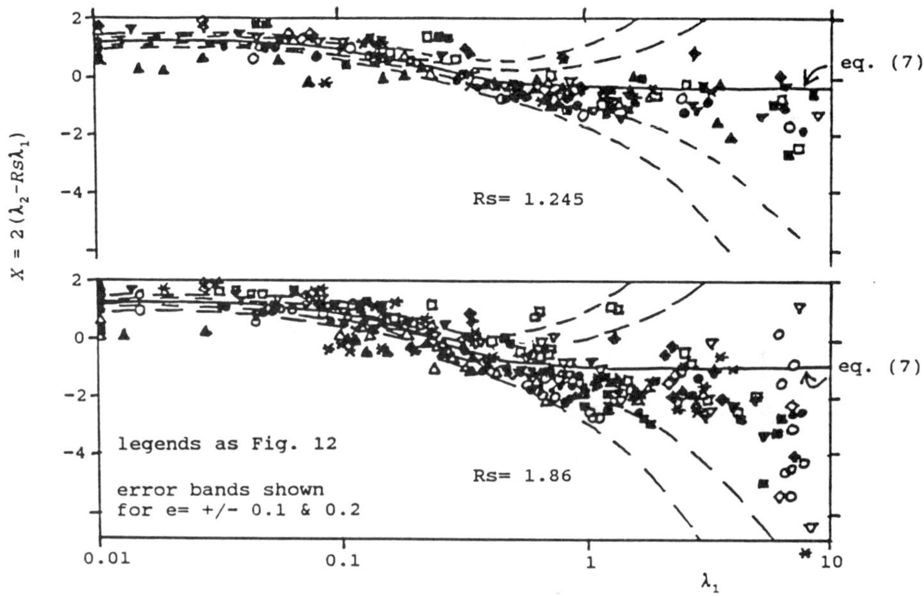

Figure 11. Plot of X vs λ_1 for Rs = 1.245 and 1.86

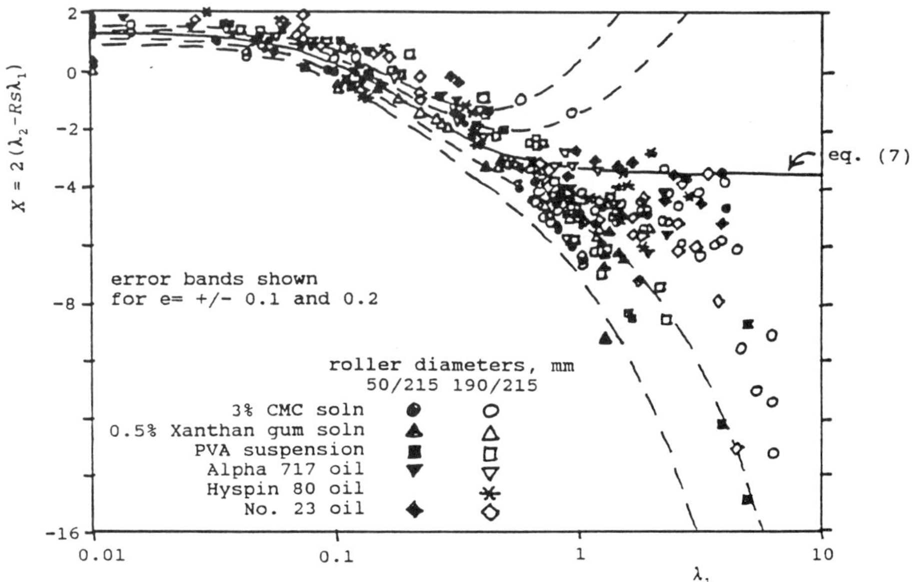

Figure 12. Plot of X vs λ_1 for Rs = 3.72

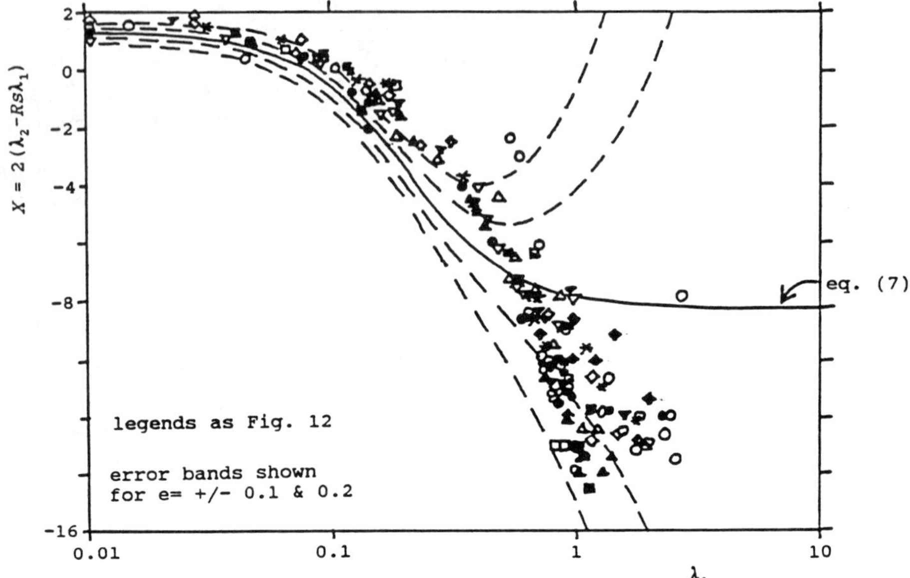

Figure 13. Plot of X vs λ_1 for Rs = 7.44

MathCAD calculations

$$x := 0,1\ ..3 \qquad Rs(x) := 2^{x}$$

$$y := -2,-1.8\ ..1.2 \qquad \lambda 1(y) := 10^{y}$$

$$Xo := 1.28$$

$$Xoo(x) := (1.28\cdot|1 - Rs(x)|)^{1.3}$$

$$X(x,y) := \frac{Xo + Xoo(x)}{1 + 1.9\cdot\lambda 1(y)^{3}} - Xoo(x)$$

2

X(x,y)

-16

0.01 λ1(y) 10

Rs(0) = 1

Rs(1) = 2

Rs(2) = 4

Rs(3) = 8

Figure 14. Improved Correlation Equation for Reverse Rolling Film Thickness

is valid only for small Rs and large λ_1. The details of deviation from theory depend on λ_1. When λ_1 is small the deviation gave a more or less constant X of X_o =1.28. When λ_1 is large, and Rs also, instead of taking the theoretical value, $X_\infty = Q^-(1-Rs)$, X dipped progressively from X_∞. The other conclusion is that X was rather insensitive to non-Newtonian properties.

To compare with previous experimental results: Benkreira et al.[2] found that λ_2 was as given by theory, ie X = 1.26(1-Rs), with λ_1 taken to be zero. This does not agree with the present results, for which we expect X to be constant at X_o for λ_1=0. Perhaps the scraping was not as effective as supposed. With reference to Figs 4-6, the difference between the constant X_o and the theoretical line is not large for Rs < 1 and it may not require much of λ_1 for the results to appear to agree with theory. On the other hand, that the non-Newtonian results were not greatly different from the Newtonian results is in agreement with the present work.

Greener and Middleman[3] did not explicitly report their results in relation to λ_1, but from their experimental arrangement, it was not zero. For low values of Rs up to unity their results agreed with theory; this is consistent with λ_1 being large. However, they found deviation at Rs= 1.75 which is equivalent to X < 1.226(1-Rs) and readily expected if λ_1 was large. Both these conclusions are in agreement with the present work. The same also goes for the relative insensitivity to non-Newtonian characteristics of the fluids used.

The results of Jurewicz[1] showed deviation from theory as Rs increased beyond unity. The curves passed through minima but the variation was not large, giving more or less constant X when taken in a broad sense. This is in accordance with the present work and suggests that the scraping was effective here and λ_1 was zero. What is very interesting about Jurewicz's results are the details of variation with U_1 and liquid viscosity. His experiments must have been very careful which more recent work has not reproduced.

Practical Use of Results

There is general agreement between the present results with previous work and as the present work covered a wider range of variables, we can look to practical applications of the correlation equation, eq. (7). Certainly it is capable of reproducing the experimental data to within about +/- 10% or 20% and we can expect the same level of accuracy in prediction. However, for large Rs and λ_1, there is a definite deviation from eq. (7) (Figs 12 and 13). A more accurate correlation is therefore desirable.

The search for a better correlation was carried out with the aid of MathCAD. The result is shown on the printout

Fig. 14. The new correlation is:

$$X = \frac{X_o + X_\infty}{1 + 1.9\lambda_1^3} - X_\infty \qquad (14)$$

where

$$X_o = 1.28 \qquad X_\infty = 1.28|1-Rs|^{1.3} \qquad (15)$$

The equation is obtained by trial and error and it is by no means the optimal best fit. However, it describes the dip in X not given by eq. (7) and can be the basis of further refinement. However, one should ask the question whether it is worth doing this or should one aim for better new experiments to obtain more precise results, as Jurewicz showed to be possible.

Implication for Theoretical Modelling

Previous theoretical work[3,4] made specific assumptions in order to obtain particular values for Q^-. The assumptions are reasonable enough each in its own right, but it is difficult to see, without appealing to experiments, how one can be favoured over the other. As Cheng[5] showed, the different assumptions that can be made is legion. It is also possible that different assumptions are valid under different speed and nip flooding conditions.

It may well be that reverse rolling is too complicated to be amenable to accurate theoretical modelling based on the one-dimensional Reynolds lubrication equation and we have to resort to numerical solutions using finite element methods. Such an approach has been used by different workers[6-15], many applying it to flows with a free surface. It is outside the scope of the present paper to pursue this further.

Whatever theoretical approach is used, it is clear that the present experimental results provide some clues to what one should consider if the theory is to be improved. For example, liquid inertia has an important role. As pointed out by Jurewicz[1], this has the effect that liquid coming through the nip and meeting a "dry" roller 1 (λ_1=0) is not dragged back into the nip as much and this gives a higher flux than fully-developed flow, ie larger λ_2 and X. In contrast, with a thick film on roller 1, large λ_1, and at high U_1, this film can be expected to be carried into the nip rather than being turned round to go onto roller 2. The film thickness on roller 2, and hence λ_2 and X, would be less than expected. The insensitivity of the results to the non-Newtonian characteristics of the fluid seems also to provide another indication of the dominant importance of inertia. Such consideration would need to be incorporated into any refinement of the theory.

5 CONCLUDING REMARKS

The present work is essentially an experimental study of reverse rolling film thickness. It is shown that theoretical modelling of the fully developed flow is inadequate to explain the complicated dependence of the film thickness on speed ratio and film thickness on roller 1 carrying fluid back into the nip. On the other hand, the results appeared to be quite insensitive to the non-Newtonian characteristics of the fluids. It seems that the results can be improved upon in terms of precision if much better care is taken in the experiments as suggested by the work of Jurewicz[1]. Nevertheless, the present conclusions have important implications for the further development of the theory. In the meantime, an empirical correlation is offered for practical application in predicting film thickness in industry.

ACKNOWLEDGEMENT

This paper is based on research done at Warren Spring Laboratory, Stevenage, as part of the work of the Co-operative Programme on Thick Liquids and Pastes: Processing and Flow (1976-84), which was funded jointly by the Department of Trade and Industry and a number of industrial companies. The Laboratory's permission for publication is gratefully acknowledged.

REFERENCES

1. J T Jurewicz, Proc Heat Transf Fluid Mech Inst, 1978, 103-116.
2. H Benkreira, M F Edwards and W L Wilkinson, Chem Engng Sci, 1981, 36, 429-434.
3. J Greener and S Middleman, Ind Eng Chem Fundam, 1981, 20, 63-66.
4. H Benkreira, M F Edwards and W L Wilkinson, Chem Engng Sci, 1982, 37, 277-282.
5. D C-H Cheng, 'Transfer of Fluids to and from Rollers I. Film Thickness in Forward Rolling of Newtonian Liquids'. Rept No. LR 505 (MH). Stevenage: Warren Spring Laboratory, 1985.
6. S F Kistler and L E Scriven, in J R A Pearson and S M Richardson (Eds), 'Computational Analysis of Polymer Processing', pp. 243-299. London: Applied Science Publishers, 1983.
7. N E Bixler and L E Scriven, IEC Res, 1987, 26, 475-483.
8. A N Beris, R C Armstrong and R A Brown, J Non-N Fluid Mech, 1986, 19, 323-347.
9. D J Coyle, C W Macosko and L E Scriven, J Non-N Fluid Mech, 1986, 171, 183-207.
10. P Townsend and E O A Carew, J Non-N Fluid Mech, 1986, 20, 293-297.
11. A Wu and S Whitaker, J Coll Interf Sci, 1986, 110, 389-397.
12. T S Obaid and P Townsend, Rheol Acta, 1985, 24, 260-264; 1984, 23, 255-260.

13. D J Coyle, C W Macosko and L E Scriven, in T Provder (Ed), 'Computer Applications in Applied Polymer Science', pp. 251-264. Washington DC: Amer Chem Soc, 1982.
14. P L Cerro and L E Scriven, IEC Fundam, 1980, 19, 40-50.
15. B G Higgins and L E Scriven, IEC Fundam, 1979, 18, 208-215.

Optimal Design of Distribution Chambers for Coating Facilities

F. Durst, U. Lange, and H. Raszillier
LEHRSTUHL FÜR STRÖMUNGSMECHANIK DER UNIVERSITÄT ERLANGEN-NÜRNBERG, D-8520 ERLANGEN, GERMANY

1 INTRODUCTION

The quality of most coated products strongly depends on the uniformity of the coated layers. For some applications film-thickness variations, even in the order of fractions of microns, are unacceptable. Because of this considerable efforts are undertaken in the coating industry to provide a constant thickness of the coated surfaces. Considering the large widths of modern coating facilities makes clear why particular efforts are needed to achieve the uniform thickness of coated films required in most applications.

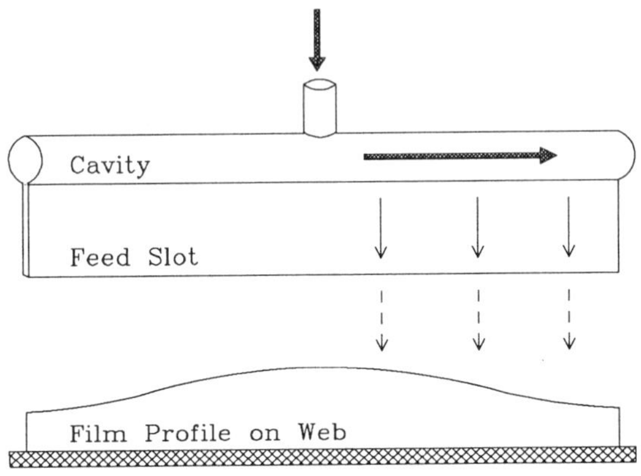

Figure 1: Uneven film thickness caused by a badly designed distribution chamber

One major reason for uneven film-thickness-distributions transversely to the direction of web motion is, that, due to poor design, the distribution cavity and the feed slot of the coating facility fail to level out the transverse pressure gradient that results from the flow within the cavity. This results in a higher pressure drop across the feed slot, and therefore in a higher flow rate in the center of the coater, where the coating liquid is supplied. Fig. 1 shows the typical "frown"-profile of the wet film caused by a badly designed distribution chamber.

The most widespread concepts for practically applied distribution chamber designs are:

- the "infinite cavity" - design, based on the idea of minimizing the film-thickness-variations by keeping the pressure loss within the cavity small compared to the pressure loss within the feed slot;
- the "coat-hanger" - design, based on the idea of matching the length of the feed slot at each location in the distribution chamber to the local pressure difference across the feed slot there, thus maintaining a constant pressure gradient across the feed slot and therefore a constant volume flux across the entire production width.

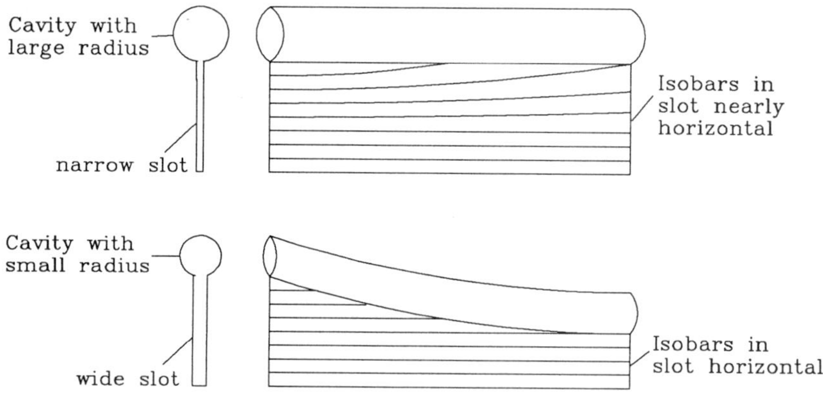

Figure 2: "Infinite Cavity" and "Coat-Hanger" Design

The literature states several different approaches to predict the performance of distribution chamber designs by simple models. Recently, Gutoff [1] presented a model for the design of "coat-hanger" chambers for shear-dependent liquids. However, Gutoff's treatment of the problem neglects the inertial effects on the fluid motion within the cavity. In some cases, inertia can have a considerable influence, as was shown by Leonard [3] for an "infinite cavity" design.

Sartor [2] presented a comprehensive model, that includes the influence of inertia, dependence of the fluid viscosity on the shear in the flow and the two-dimensional flow within the feed slot. However, the complexity of Sartor's modeling equations, which have to be solved by a two-dimensional finite element procedure, appears to make this model unsuitable for studies of large parameter ranges, as well as for a qualitative analysis for the flow through the feed slot, which could clarify the important mechanisms of the distribution flow.

In this paper, a model equation for the flow in distribution chambers is proposed, which is as comprehensive as Sartor's model, but reduces the computational effort to the solution of a single ordinary differential equation by assuming a one-dimensional flow in the feed slot. This causes, at least for the chamber geometries relevant in the applications, only a small loss of accuracy.

The model equation is suited for a qualitative analysis. Its solutions predict film-thickness-variations for a given distribution chamber design in dependence on the desired wet film thickness, the web velocity and the rheological properties of the coating liquid. On the other hand, solving the so-called inverse problem of this equation leads to a distribution chamber design, for which uniform film thickness can be expected at a given operating point. These two different interpretations of the model equation yield an optimization strategy on which a distribution chamber design for specific coating applications can be found.

Moreover, the model presented does not treat cavities with non-circular cross sections in the usual hydraulic-radius-approximation (cf. [1]). This approximation would lead to considerable errors for geometries used in practice, as, for instance, cavities with half-circular or "tear-drop" cross-sections (cf. [2]). The model equations derived in this paper are such that, in order to treat these cavities with the same accuracy as circular ones, only three constants, which codify the shape of the cross section of the cavity have to be determined in advance, either numerically, or, in some cases, analytically, with a computer algebra system like MAPLE, as was done for this work.

2 DESCRIPTION OF THE MODEL

Similar to Sartor's approach, the model presented in this paper is based on averaged mass and momentum balances within the cavity, which are obtained by integrating the flow equations over the cross section of the cavity. This yields the following mass balance:

$$\frac{d\dot{Q}}{ds} = \dot{q}(s) \quad . \tag{1}$$

In this equation, $\dot{Q}(s)$ means the flow rate through the cavity cross section at s, which is the curvilinear coordinate measured along the cavity axis. The local flow rate per unit length from the cavity into the feed slot is denoted by $\dot{q}(s)$.

Averaging the momentum equations leads to the following relation between pressure drop and flow rate in a cavity:

$$\frac{dp_K}{ds} = -\frac{C_2}{|A|^2}\mu_Q\dot{Q} + C_3\frac{1}{|A|}\left(\mu_Q\frac{d^2\dot{Q}}{ds^2} + 2\frac{d\mu_Q}{ds}\frac{d\dot{Q}}{ds}\right) - C_1\frac{\rho}{|A|^2}\dot{Q}\frac{d\dot{Q}}{ds} \quad . \tag{2}$$

Here, p_K is the pressure, averaged over the cross sectional area $|A|$ of the cavity. The density of the fluid is denoted by ρ, while μ_Q, which codifies the dependence of the fluid viscosity on the flow rate, is given by

$$\mu_Q(s) = m\left(\frac{\dot{Q}(s)}{|A|^{3/2}}\right)^{n-1} \tag{3}$$

with m resp. n being the consistency resp. the power-law index of the Ostwald-de Waele model for the shear dependence of the fluid viscosity.

The dimensionless constants C_1, C_2, and C_3 are given by

$$C_1 := \frac{1}{|A|}\int_A f^2 dA \ , \tag{4}$$

$$C_2 := -\int_A \frac{\partial}{\partial n_1}\left(\mu_f\frac{\partial f}{\partial n_1}\right) + \frac{\partial}{\partial y}\left(\mu_f\frac{\partial f}{\partial y}\right) dA \ , \tag{5}$$

$$C_3 := \frac{1}{|A|}\int_A \mu_f \cdot f dA \ . \tag{6}$$

In here, $f(n_1, n_2)$ is the velocity profile for fully developed flow in a duct with the same cross section as the cavity, which is normalized to the mean velocity, and μ_f denotes

$$\mu_f = \left(|A|\left(\left(\frac{\partial f}{\partial n_1}\right)^2 + \left(\frac{\partial f}{\partial n_2}\right)^2\right)\right)^{(n-1)/2} \quad . \tag{7}$$

The coordinates n_1 and n_2 are perpendicular to each other as well as to the duct axis.

The constants C_i can be either determined analytically for circular and other simple cross sections, or computed numerically for complex cross sections. In this paper, not only the circular case is considered, but also half circles and circular segments ("tear-drops"). According to Moffatt and Duffy [4] the normalized velocity profile for fully developed newtonian flow in a duct, whose cross section is a circular segment with opening angle 2α and radius R, is given by

$$f(r,\theta) = -G\left(\frac{1}{4}\left(\frac{r}{R}\right)^2\left(\frac{\cos(2\theta)}{\cos(2\alpha)} - 1\right) - \sum_{n=0}^{\infty}\frac{2\cdot(-1)^n}{\alpha\lambda_n(\lambda_n^2-4)}\left(\frac{r}{R}\right)^{\lambda_n}\cos(\lambda_n\theta)\right) . \tag{8}$$

In this equation, (r, θ) is a polar coordinate system with origin in the corner of the circular segment. The constants λ_n and G are defined as

$$\lambda_n = \frac{\pi}{2\alpha}(2n+1) \ . \tag{9}$$

$$G = \left(\frac{1}{16\alpha}(2\alpha - \tan(2\alpha)) + \frac{4}{\alpha^2}\sum_{n=0}^{\infty}\frac{1}{\lambda_n^2(\lambda_n^2+2)^2(\lambda_n-2)}\right)^{-1} \tag{10}$$

The formulae for the constants C_i with f given by (8) can be easily simplified with the help of a symbolic computer code:

$$C_1 = \frac{2G^2}{\alpha}\left(\frac{1}{96}\left(\frac{\alpha}{2}\left(2+\frac{1}{\cos^2(2\alpha)}\right)-\frac{3}{4}\tan(2\alpha)\right)+ + \sum_{n=0}^{\infty}\frac{3}{\alpha\lambda_n^2(\lambda_n^2-4)^2(\lambda_n+4)(\lambda_n+1)}\right) \quad (11)$$

$$C_2 = -G\cdot\alpha \quad (12)$$

$$C_3 = 1 \quad (13)$$

These formulae can be evaluated very effectivey, since the occurring series converge very fast.

If one assumes a one-dimensional flow in the slot and sets the outer pressure to zero, the local pressure $p_K(s)$ within the cavity can be easily expressed in terms of the local slot length $h(s)$ and the slot width B, which, for simplicity, is assumed to be constant here:

$$p_K(s) = m\cdot\left(\frac{\dot{q}(s)}{2}\right)^n\left(\frac{2n+1}{n}\right)^n B^{-(2n+1)}h(s)\ . \quad (14)$$

Eliminating $\dot{q}$ and p_K from equations (1),(2) and (14) yields the model equation stated below in dimensionless quantities.

$$K_3\left(\frac{dh^*}{dx^*}\frac{d\dot{Q}^*}{dx^*}+nh^*\frac{d^2\dot{Q}^*}{dx^{*2}}\right)\left(-\frac{d\dot{Q}^*}{dx^*}\right)^{n-1} = s_{x^*}(\dot{Q}^*)^n - -K_1\left[(\dot{Q}^*)^{n-1}\left[\frac{d^2\dot{Q}^*}{dx^{*2}}s_{x^*}^{-1} - \frac{s_{x^*x^*}}{s_{x^*}^2}\frac{d\dot{Q}^*}{dx^*}\right]+2(n-1)(\dot{Q}^*)^{n-2}\left(\frac{d\dot{Q}^*}{dx^*}\right)^2 s_{x^*}^{-1}\right]+ + K_2\cdot\dot{Q}^*\cdot\frac{d\dot{Q}^*}{dx^*}\ , \quad (15)$$

In this equation, $\dot{Q}^*$ is the flow rate with the cavity, non-dimensionalized by the total flow rate, and h^* is the slot length, non-dimensionalized by the half of the production width L. Moreover, the curvilinear coordinate s was replaced by the cartesian coordinate x^*, which is non-dimensionalized L. s_{x^*} and $s_{x^*x^*}$ are the first and second derivatives of the arc-length s with respect to x^*:

$$s_{x^*} = \sqrt{1+\left(\frac{dh^*}{dx^*}\right)^2} \quad (16)$$

$$s_{x^*x^*} = \frac{d^2h^*}{dx^{*2}}\frac{dh^*}{dx^*}\left(\sqrt{1+\left(\frac{dh^*}{dx^*}\right)^2}\right)^{-1}\ . \quad (17)$$

The dimensionless constants K_i are given by

$$K_1 = \frac{C_3}{C_2}\cdot\left(\frac{|A|}{L^2}\right)\ , \quad (18)$$

$$K_2 = \frac{C_1}{C_2}\cdot Re\ , \quad (19)$$

where Re is the Reynolds number, and

$$K_3 = \frac{1}{C_2}\left(\frac{|A|}{L^2}\right)^{(3n+1)/2}\left(\frac{L}{B}\right)^{2n+1}\left(\frac{2n+1}{2n}\right)^n \quad . \tag{20}$$

3 RESULTS

Optimization Strategy

There are two possible interpretations of the model equation. On one hand, it can be taken as an ordinary differential equation for the flow rate $\dot{Q}^*$, which can be solved for a **given** slot shape $h^*(x^*)$ subject to the boundary conditions

$$\dot{Q}^*(0) = 1 + \beta \ , \qquad \dot{Q}^*(1) = \beta \quad . \tag{21}$$

These conditions allow for a bypass rate β, i.e. a part of the total flow rate, which does not leave the cavity through the feed slot, but through a sink at the end of the cavity. Bypasses are often used in practice to prevent shear-thinning suspensions from sedimentation. Thus the model can be used to predict relative film thickness variations for a fixed distribution chamber design.

On the other hand, the model equation can be interpreted as an ordinary differential equation for the determination of the **slot shape** $h^*(x^*)$, such that for a given operating point, determined by the values of the web velocity U_B, the desired film thickness H_F and the fluid properties m, n and ρ, it leads to constant wet film thickness, which implies that Q^* is put equal $1 + \beta - x^*$. In this inverse form, the equation can be used to optimize the chamber design for a given operating point.

Thus, for a given operating range, i.e. ranges for the above mentioned parameters, the optimal chamber design can be determined by the following strategy:

1. Choose values for the cross sectional area of the cavity and for the slot width. Those values depend on the admissible range of shear rates within the chamber, which are predetermined by the rheological properties of the coating liquid.

2. Choose a point in the middle of the operating range and determine the appropriate slot shape for uniform outflow $\dot{q} = 1$.

3. Extend the slot length to its maximum value, which is predetermined by the highest admissible pressure in the chamber. Calculate the maximum deviations of wet film thickness in the operating range.

4. If the calculated wet film thickness deviations are tolerable, accept the chamber design. If they are not acceptable, modify the cavity's cross section and the slot width and repeat the above procedure, starting with step 2.

<u>Qualitative Analysis</u>

From equation (refpro3), it is easily seen that a uniform flow rate q^* in the slot can only be achieved if the feed slot length $h(x)$ is shaped proportional to the pressure p_K within the cavity. This is the basic idea of the coat-hanger strategy. As stated above, the appropriate shape of a coat-hanger distribution chamber of a coater can be found by writing the model equation (8) for the flow rate $\dot{Q}^* = 1 + \beta - x^*$ in the cavity:

$$K_3 \frac{dh^*}{dx^*} = K_2 \cdot (1 + \beta - x^*) - \sqrt{1 + \left(\frac{dh^*}{dx^*}\right)^2} (1 + \beta - x^*)^n \ . \tag{22}$$

For simplicity, the K_1-term was neglected in this equation, since, for realistic chamber geometries, it is insignificant compared to the other terms. As for small values of K_2, the solution of (22) is expanded into an asymptotic series:

$$h^* = h_0^* + K_2 \cdot h_1^* + O(K_2^2) \quad , \text{as } K_2 \to 0 \ . \tag{23}$$

At zeroth order, insertion of this series into (22) results in:

$$K_3 \frac{dh_0^*}{dx^*} = -\sqrt{1 + \left(\frac{dh_0^*}{dx^*}\right)^2} (1 + \beta - x^*)^n \ . \tag{24}$$

This shows that the appropriate slot shape decreases with the distance from the center of the distribution chamber (where the inlet is located), as is expected from equation (14). Derivation of (24) gives an expression for the curvature of the slot shape:

$$\frac{\dfrac{d^2 h_0^*}{dx^{*2}}}{\left(1 + \left(\dfrac{dh^0}{dx^*}\right)^2\right)^{3/2}} = \frac{n}{K_3} (1 + \beta - x^*)^{n-1} \ . \tag{25}$$

Thus, the slot shape curvature is always positive, as shown in fig. 2. For newtonian liquids ($n = 1$), the curvature of the feed slot is constant. For shear-thinning liquids ($n < 1$), the curvature increases with the distance from the center of the distribution chamber. This is caused by lower shear rates, and therefore higher viscosities at the end of the cavity. If the bypass rate β becomes zero, the curvature at the end of the cavity becomes infinite, since the Ostwald-de Waele model for shear-thinning liquids predicts infinite viscosity, and therefore meaningless results, at shear rate zero.

The first order term of the asymptotic expansion is given by:

$$\frac{dh_1^*}{dx^*} = \frac{K_3}{K_3^2 - (1 + \beta - x^*)^{2n-1}} \ . \tag{26}$$

This means that, since $K_3 > 0$ for any reasonable geometry, the slot shape becomes less steep as the Reynolds number increases. This is due to the fact that the flow in the cavity is decelerated because of the decrease of the flow rate. The inertial force acts against this deceleration, which results in a smoother pressure gradient across the cavity, and therefore in a more levelled slot shape.

The asymptotic analysis of (22) for small K_2 illustrates the influence of shear-dependence and inertia on the appropriate coat-hanger shape for uniform outflow. An asymptotic expansion of (22) for $K_3 \to \infty$ shows how the infinite cavity concept is related to the coat hanger design, resulting in:

$$\frac{dh^*}{dx^*} = O(K_3^{-1}) \quad , \text{as } K_3 \to \infty \quad , \tag{27}$$

which means that a uniform slot length $h^*(x^*) = H^*$ is acceptable for large values of K_3. Equation (20) shows

$$K_3 \propto \frac{|A|^{1/2}}{B} \cdot \left(\frac{|A|^{3/2}}{LB^2}\right)^n \quad . \tag{28}$$

Thus, K_3 indeed is large for cavities with large cross sections and narrow feed slots, which are the main design features of an "infinite cavity". On the other hand, (28) indicates that an infinite cavity is not well suited for strongly shear-thinning liquids, since K_3 decreases with the decrease of n.

Computational Results

Slot shapes h^* were computed numerically from (22) using a Runge-Kutta solver for ordinary differential equations. Equation (22) was derived from the model equation (15) by neglect of the small K_1-term. This means the loss of the highest derivative of h^*. Nevertheless, this is no singular perturbation problem, since the boundary conditions can still be satisfied by solutions of (22).

Film thickness variations and pressure distributions for given slot designs were calculated from the model equation (15) with boundary conditions (21) by linearizing the equation in the vicinity of the desired cavity flow rate $\dot{Q}^* = 1 - x^*$ and solving the resulting linear boundary value problem by a shooting method based on the Runge-Kutta solver. This took less computation time than solvers for boundary value problems would have taken, since no systems of equation had to be solved. The linearization, which is equivalent to one single Newton-iteration step, yielded in no significant loss of accuracy, since deviations of a few percent from the desired flow rate were considered intolerable for the chamber design.

Fig. 3 shows the influence of the inertial force on the appropriate slot shape. As the figure shows, the slot shape becomes less steep as the Reynolds number increases, which is in accordance with the qualitative considerations above. Fig. 4 and 5 show the deviations of the wet film thickness and the corresponding cavity pressures, predicted by the model for different Reynolds numbers, if the slot shape was designed without taking into account the inertial effects.

Fig. 4 points out, that the inclusion of the inertial effects into the treatment of the chamber flow is essential, as is the use of the exact solutions for non-circular cross sections. Fig. 6 and 7 show the appropriate slot shapes for half-circular cavities at $Re = 0$ and $Re = 5$, which were predicted using the common hydraulic-radius-approximation on one hand, and Moffatt's and Duffy's exact solution on the other hand. The different results are caused by the fact that the hydraulic-radius-approximation overestimates the viscous pressure drop, and underestimates the inertial effects in the half-circle shaped cavity. This may result in large deviations of the film thickness across the production width, as shown in fig. 8 for $Re = 0$.

Finally, fig. 9 shows the predicted slot shapes for different shear-thinning liquids. Due to low viscosities near the center of the cavity and high viscosities at its end, the slot shapes become less curved with decreasing the power-law index. On the other hand, the slots become steeper with a decreasing power-law index, since the pressure drop across the feed slot diminishes due to the high shear rates and therefore low viscosities within the slot. In the calculations of the slot shapes shown in fig. 9, a bypass rate of 10 percent was assumed.

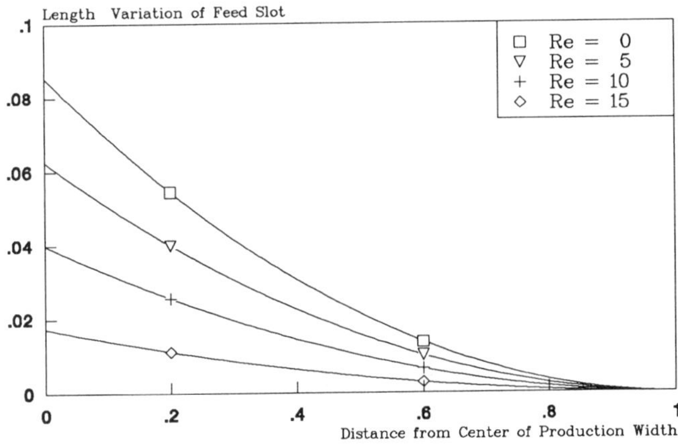

Figure 3: Slot shapes for different Reynolds numbers

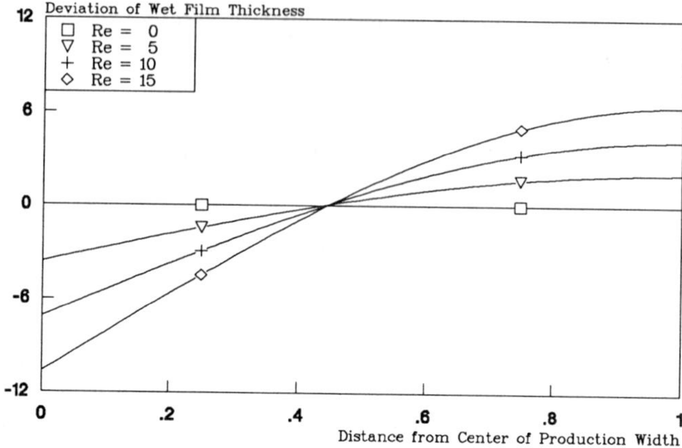

Figure 4: Film thickness deviations for different Reynolds numbers in a chamber designed for $Re = 0$

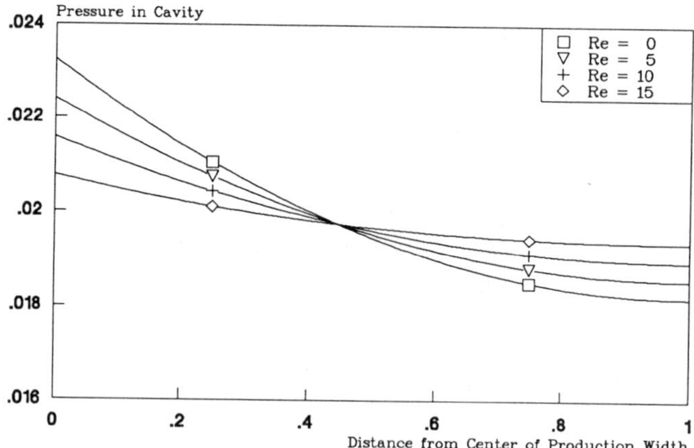

Figure 5: Cavity pressure for different Reynolds numbers in a chamber designed for $Re = 0$

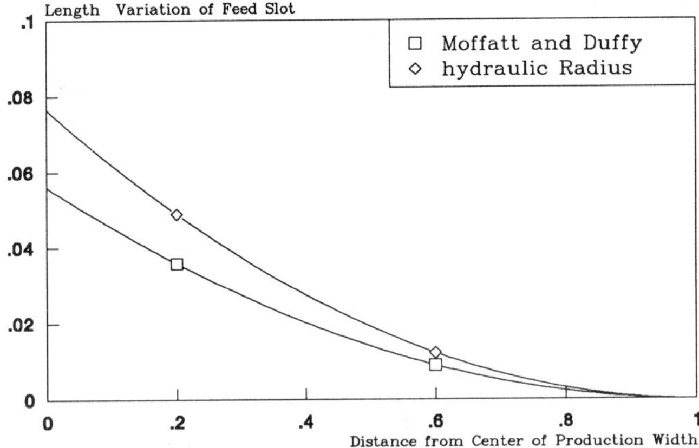

Figure 6: Predicted slot shapes for a half-circular cavity using the hydraulic-radius-approximation compared to Moffatt and Duffy's solution at $Re = 0$

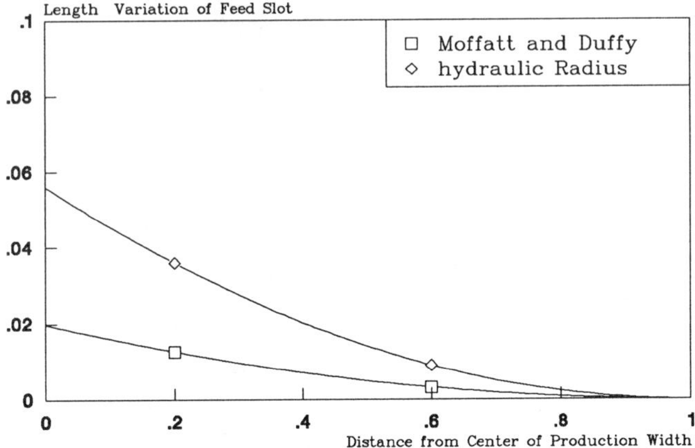

Figure 7: Predicted slot shapes using the hydraulic-radius-approximation or Moffatt and Duffy's solution at $Re = 5$

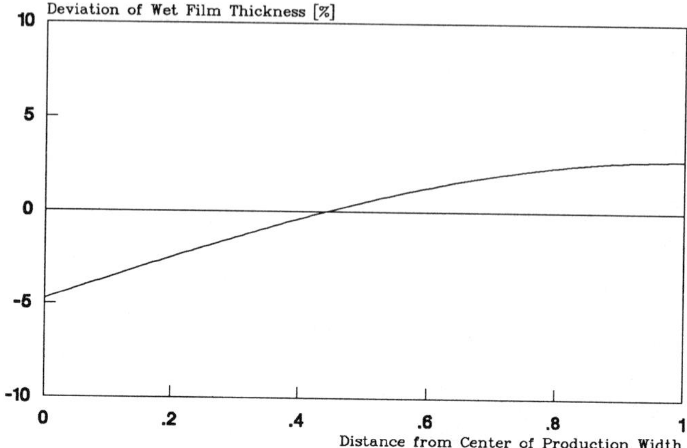

Figure 8: Film thickness deviations in a half-circular chamber designed using the hydraulic-radius-approximation

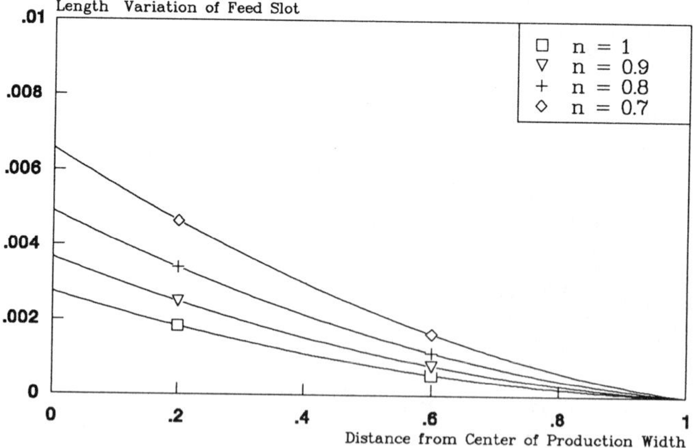

Figure 9: Slot shapes for different shear-thinning liquids at a bypass rate of 10 percent

4 CONCLUSION

The model for flow in distribution chambers presented in this paper allows for both the determination of the appropriate chamber design at a given operating point and the prediction of wet film thickness variations for given chamber designs. A combination of these two computational steps yields a strategy for finding the optimal chamber design for a given range of operating parameters.

The qualitative analysis of the model equation clarifies the ideas of the "coat-hanger" and the "infinite cavity" chamber design concepts. It also illustrates the influence of shear-dependence and inertia on the appropriate feed slot shape.

The computed results show how important it is to include inertial effects and shear-dependence in the model equations. In case of non-circular cavities, it is pointed out that the hydraulic-radius-approximation which is used in most chamber design models may lead to considerable errors. These are avoided by the presented model, which codifies the influence of the cross section in the form of three constants. These constants can be calculated in advance by a numerical procedure or by computer algebra.

The use of the Ostwald-de Waele model for shear-thinning liquids leads to meaningless results if the shear rate becomes zero at the end of the cavity. In a cavity with a bypass, used in practice to avoid too low shear rates at the end of the production width, this problem does not occur. On the other hand, even in the case with bypass the simple Ostwald-de Waele model provides only a qualitative insight on the influence of shear-dependence.Inclusion of a better model for rheological properties therefore would be a very important step in the development of the model into a comprehensive tool for the optimal chamber design. Another important refinement would be the extension of the model to the flow situation in tapered cavities, which are used in practice to level the shear rate variations within the chamber.

REFERENCES

[1] Gutoff, E.B., Simplified Design of Coating chamber Internals, presented at AIChE Annual Spring Meeting, New Orleans, 1992

[2] Sartor, L., Slot Coating: Fluid Mechanics and chamber Design, Ph. D. Thesis, University of Minnesota, 1989.

[3] Leonard, W. K., Inertia and Gravitational Effects in Extrusion Dies for Non-Newtonian Fluids, Polymer. Eng. and Sci., vol. 25, no. 9 (1985), pp. 570-576.

[4] Moffatt, H.K. and Duffy, B.R., Local Similarity Solutions and Their Limitations, J. Fluid Mech. vol. 96 part 2 (1980), pp. 299-313

Thin Films With Direct Gravure Coating

Hadj Benkreira
DEPARTMENT OF CHEMICAL ENGINEERING, UNIVERSITY OF BRADFORD, BRADFORD BD7 1DP, UK

INTRODUCTION

Many coating applications require the production at high substrate speeds (> 1m/s) of very thin, uniform films (4-20 microns wet gauge). Typical examples include audio, video and computer tapes, photographic films and a variety of flexible packaging substrates. Gravure coating is one of the few coating processes capable of achieving such thin gauges at high speeds. As the name implies, the operation uses a knurled steel roller with a pattern which is either chemically or mechanically engraved on it. Typical patterns (see Figure 1) are the quadrangular, trihelical and

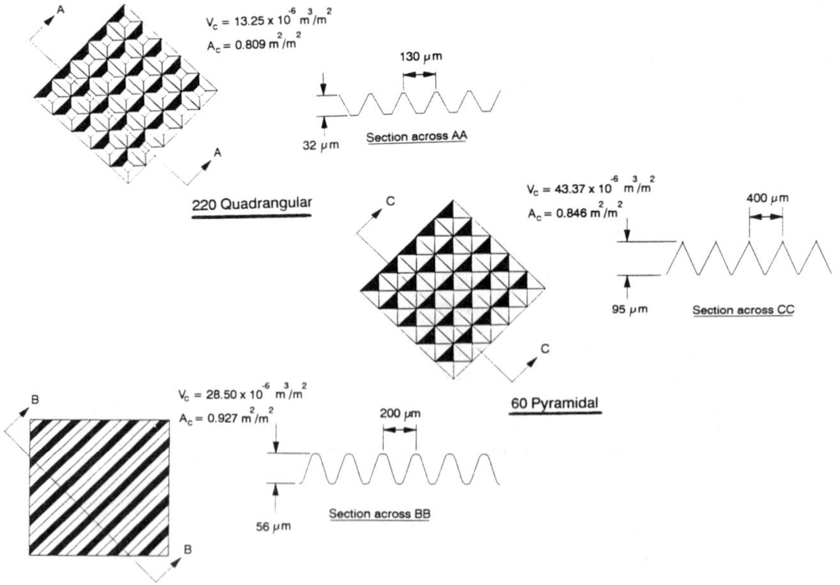

Figure 1: Schematic representation of typical gravure geometries

pyramidal with cell volume per unit area of roller surface of order 10-50x10^{-6} m^3/m^2 and a wetted area coverage of 0.80 - 0.90 m^2/m^2 of roller surface. These cells are flooded with liquid and wiped by a blade pressed against the rotating roller. The ensuing liquid is then transferred directly onto the substrate (direct gravure, Figure 2a) or onto a transfer roll which applies the coating to the substrate (indirect or offset gravure coating, Figure 2b).

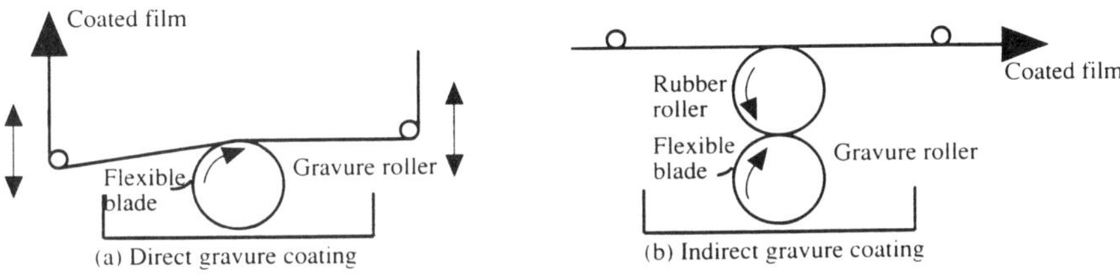

Figure 2: Two modes of gravure roll coating

The applied blade load and the hydrodynamic load generated by the liquid underneath the blade control the extent of wiping, ie the thickness of the film formed over the filled gravure cells. Clearly, where thin coatings are required, gravure roll coating must be used in situations where the hydrodynamic load under the blades are small so that the blade essentially wipe clean the periphery of the roller leaving only the gravure cells filled with liquid. At high coating speeds, this is possible only with low viscosity liquids or rigid blades (these damage the cell geometry).

This paper examines the fluid mechanics of the direct gravure coating of low viscosity liquids. Here the geometry of the cells and the operating variables such as ratio of substrate over the roller speeds and the viscosity of the liquid become the important variables. It is important to note that despite its wide usage, this operation has received little research interest [1-3].

EXPERIMENTAL METHOD

A gravure roll coating rig complete with substrate handling facilities was used for this study. Three chromium plated gravure rollers, 220 quadrangular, 85 trihelical and 60 pyramidal with cell volume per unit area of roller surface, V_c, values of 13.25 x 10^{-6}, 28.50 x 10^{-6} and 43.37 x 10^{-6} m^3/m^2 respectively were used (see Fig.2). The pre-fixes 220, 85 and 60 refer to the number of cells or "lines per inch" on the roller surface. The projected areas (wetted exposed surfaces), A_c, of these rollers were 0.809, 0.927 and 0.846 m^2/m^2 roller surface respectively. The gravure rollers, 0.1 m in diameter and 0.182 m wide, and the substrate wind-up roller speed were driven by remote controlled variable speed electric gearboxes (Brown Pestell SK20-PM90S and SK20-DPM40 respectively). The coating liquid was fed via a feed slot made from perspex and underneath a flexible plastic blade (250 μm thickness) of flexural rigidity 540 x 10^{-6} N/m^2 and length 9 x 10^{-3} m. Four water based Newtonian solutions, labelled A, B, C and D, with viscosities in the range 0.001 - 0.0134 Pa.s were used; their physical properties are shown in Table 1.

Test Liquid	Viscosity Ns/m^2	Surface Tension N/m	Density kg/m^3
A	0.001	0.0412	1002
B	0.0037	0.0408	1006
C	0.0057	0.0412	1004
D	0.0134	0.0415	1005

Table 1: Physical properties of test liquids

For the experiments with the speed ratio, U_s/U_r, set to 1.0, the coating speed was in the range 0.42 to 5.0 m/s. In the experiments where the speed ratio and the viscosities were varied, only the substrate speed was altered, keeping the roller speed constant at 0.83 m/s. The gravure and substrate wind-up roller speeds were measured by the use of micro switches mounted on the roller shafts and linked to a digital counter (Racal 9523). The external pressure applied on the blade was controlled by varying the slop gap, h_o, from 0.2 to 0.4 mm and the angle of wrap of the substrate, θ_s, varied from 0° to 12°. The thickness of film transferred onto the substrate, h_t, was measured[4] using a two-piece infrared thickness gauge (infra-red Engineering, model 0616) with the transmitter and receiver placed 100 mm in either side of the substrate.

ANALYSIS

The gravure coating flow is a difficult system to model theoretically; it not only involves a free surface, but withdrawal from tiny cavities interlinked in a complex geometrical arrangement. Dimensional analysis was thus considered to be a more suitable approach to assess the variation of transferred film thickness, h_t, with operating conditions, viz:

$$h_t = \left(U_r, U_s, \mu, \sigma, \rho, h_o, \theta_s, V_c / A_{c,} R, g\right) \qquad (1)$$

For operations where the roller radius (R) is kept unchanged and surface tension (σ) and gravity (g) effects are negligible, equation (1) transforms to:

$$h_t \frac{A_c}{V_c} = f\left(\frac{\rho U_s}{\mu}\frac{V_c}{A_c}, \frac{U_s}{U_r}, h_o \frac{A_c}{V_c}, \theta_s\right) \qquad (2)$$

where $h_t A_c/V_c$ is the dimensionless film thickness, $\rho U_s (V_c/A_c)/\mu$ the Reynolds number (Re), $h_o A_c/V_c$ the dimensionless slop gap, θ_s the substrate wrap angle and

U_s/U_r the substrate to roller speed ratio. The variation of the dimensionless film thickness h_tA_c/V_c with the various parameters represented in Equation 2 is now described.

RESULTS & DISCUSSION

(i) Effect of Blade Loading, h_0A_c/V_c

In true gravure roll coating such an effect can be ignored as the blade loading is designed to remove the excess liquid over the cells. This parameter was included in our analysis to ensure that the gravure roll coating was indeed reproduced in all our experiments which showed no variation in film thickness with h_0A_c/V_c in the range 3.90 to 24.42.

(ii) Effect of Substrate Wrap Angle, θ_s

Here, wrap angle in the range 0° to 12° corresponding to a number cell coverage (θ_sR/cell width) in the range 0 to 81 (220 quadrangular: 0-81,85 trihelical: 0 - 52 and 60 pyramidal: 0 - 26) led to no discernible variation in film thickness in all our experiments. This suggests that the mechanism of film transfer is controlled by drag from a small pool which is formed at the point of contact and that no pressure is developed inside the wrapped region. The experimental results are now discussed in the light of this transfer mechanism.

(iii) Effect of Reynolds number, speed ratio and Cell Geometry

Figure 3 shows the variation of the dimensionless film thickness against the Reynolds number, at speed ratio = 1.0, for the three roller geometries at all wrap angles and dimensionless slot gaps (243 data points). The dimensionless film

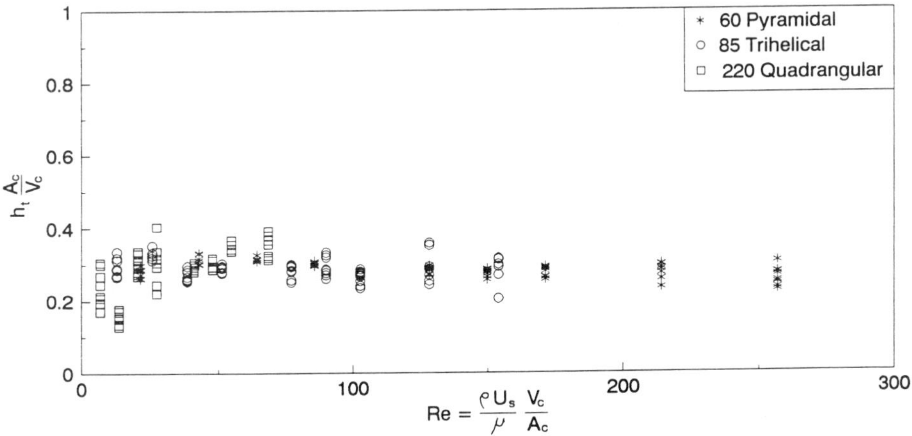

Figure 3: Variation of Dimensionless Film Thickness with Reynolds number at speed ratio =1

thickness shows a sharp increase with increasing Reynolds number below 30, a mild decrease up to Reynolds number about 75 and then attains a constant value of 0.30 +/-9%. The same behaviour is observed when the speed ratio is varied between 0.5 and 2.0 (in 0.25 increments), as shown in Figure 4 with the limiting Reynolds number

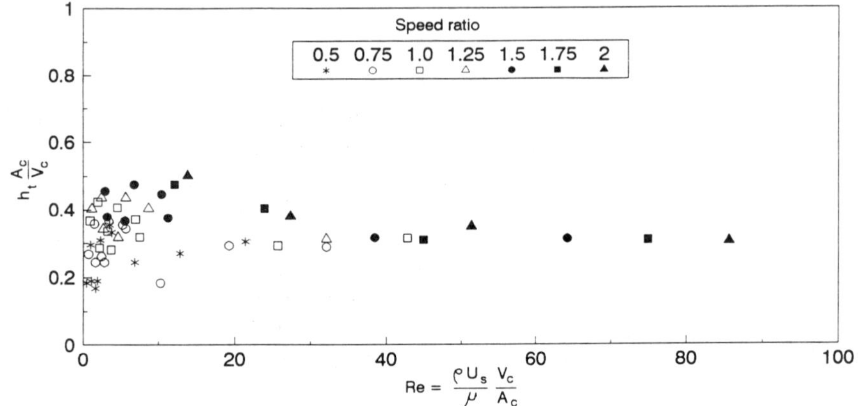

Figure 4: Variation of Dimensionless Film Thickness with Reynolds number at various speed ratios

shifting to 15 and 45 compared to 30 and 75. The same constant value of 0.30 +/- 5% is attained at higher Reynolds number.

The important and practical implication of these findings is that in normal gravure operation (large Re), the film thickness transferred to the substrate h_t, is governed by the size of the gravures only represented by V_c/A_c, in a simple manner.

$$h_t = 0.30\ V_c/A_c$$

which suggests as a rule of thumb that only 1/3 of the descriptive cell size V_c/A_c is transferred as a film. From the practical view point the observations made here are important since the choice of gravure roller to produce a desired film thickness at large Reynolds numbers can be predicted without the need for extensive trials.

The behaviour at low Reynolds number is less predictable, but grossly speaking the film thickness is observed to increase with increasing Reynolds number, i.e. with increasing speed and cell size and decreasing viscosity. It also may attain values as high as 0.5 as the speed ratio increases. These observations correspond to those made by Pulkrabek and Munter[3] but are not representative of effective gravure operation where a great degree of accuracy and good behaviour with slightly changing conditions are expected. At low Reynolds number, surface tension and gravity effects become important as highlighted by standard withdrawal coating[5-7].

Finally it is important to note that at low Reynolds number there appears to be a slight maximum in the film thickness curves for the trihelical and pyramidal cell configurations (Figure 4) suggesting that larger films (only marginally) are formed at low speeds i.e. these cells empty better at low speeds. The reverse is observed with the quadrangular geometry where a minumum occurs at low speeds. These observations are relevant in practice during the start and end of a coating run where the thickness may be expected to vary, be it only marginally.

CONCLUSIONS

Measurements of film thickness have been made for a gravure roll coater under varying operating conditions of slot gaps, coating speeds, roller roughness and substrate wrap angles. Extensive experimental data is presented and a simple but important design correlation has been established. The important conclusion that can be drawn from this study is that for most practical operating conditions (large Re), the transferred film thickness, h_t, is independent of all the operating conditions bar the gravure geometry. The amount of liquid available for transfer is controlled by the gravure size only and is about 1/3 of the cell volume per unit wetted area of roller. Deviations from this value are observed at low Reynolds numbers.

REFERENCES

1. Witt F., 1977, Reverse gravure - existing coating alternative to the air knife - part 1. *Paper, Film and Foil Converter,* **15** 41-43.

2. Statyuk I.S and Peglovskii V.L., 1981 Study of the thickness of polymer coatings applied by screen rolls. *Khim, Neft.* **9** 9.

3. Pulkrabek W.W. and Munter J.D., 1983, Knurl roll design for stable rotogravure coating. *Chem. Eng. Sci.* **38** 1309-1314.

4. Patel R., 1989. Ph.D Thesis, Univ. of Bradford, U.K.

5. Subbaraman C.V., Ph.D. Thesis, Univ. of Bradford, U.K.

6. Spiers R.P. and Wilkinson W.L., Uniqueness of film thickness and meniscus profiles in vertical withdrawal. *Chem. Eng. Sci.* **29** 1821-1825.

7. Spiers R.P., Subbaraman C.V. and Wilkinson W.L., 1974, Free coating of Newtonian fluids onto a vertical surface. *Chem. Eng. Sci.* **29** 389-396.

Mechanisms and Mechanics of Air Entrainment in Coating Processes

Roger Burley
DEPARTMENT OF CHEMICAL ENGINEERING, HERIOT-WATT UNIVERSITY, RICCARTON, EDINBURGH EH14 4AS, UK

1 ABSTRACT

Industrial processes which involve the coating of a fluid layer onto a moving solid surface will always be constrained to a finite velocity due to several causes which relate to the coating fluid and the substrate. An overview of these factors which will affect all high speed coating processes irrespective of the actual type of equipment being used has already been published[1].

This paper is concerned in more detail with the actual mechanisms through which air entrainment eventually arises as fluid forces at a three-phase junction coincide and interact. The behaviour of the unstable contact line which then occurs may be further modified by other factors such as surface roughness or charge.

The prediction of the onset of air entrainment has been the subject of many studies and typically may be represented by equations of the type which relate to the velocity at which air commences becoming entrained, V_{ae} by equations of the form[1].

$$V_{ae} = 67.68 \left[\mu \frac{g}{\emptyset \sigma}\right]^{-0.672}$$

which reduces to:

$$V_{ae} = 1.14 \left[\frac{\sigma}{\mu}\right]^{0.77}$$

for most process coating fluids of density, ϕ, viscosity μ, and interfacial tension σ, where g is the acceleration due to gravity.

Given the range of values of these variables, it is clear that interfacial tension and viscosity, or more correctly their ratio will be extremely important in predicting processing speeds. It also follows that accurate rheological data will be required for prediction of the maximum processing velocity which avoids air entrainment. The nature of coating fluids are normally determined by the end product requirement so that there will be relatively little room for manoeuvre within a specific coating operation.

The manner in which the air or second immiscible phase will always be entrained by any moving surface, is described and the air-entrainment process followed through ab initio.

The mechanisms which give rise to fairly predictable process coating velocities are described in detail and show how the air is mechanically entrained along with the moving surface which comes into contact with the coating or contacting fluid.

The entrained air takes the form of bubbles whose size depend upon fluid properties and may be seen as 'bulk' air entrainment rather than as the micron-layer always entrained adjacent to the surface asperities. It enters via an unstable three-phase wetting line and eventually coalesces into bubbles or dissolved. These micro-bubbles may detach into the coating or become encapsulated between the coating fluid and surface.

Controlling the onset of air entrainment is not always possible but the relationships above show that alteration of the coating solution viscosity may produce quite a significant increase in the velocity before bulk air entrainment occurs. It has further been recently determined that there is indeed a maximum velocity of bulk air entrainment upon which altering the interfacial tension has no effect. This was first estimated to be approximately 10.5 cm sec^{-1} by Kennedy[7] but more recently refined by O'Connell[1,3] to 8.9 cm sec^{-1}.

This study looks at the detailed description of bulk air entrainment at the three point contact line up to and as the dynamic contact angle, δ_d tends to 180°. At this point the entrained air instantaneously and repeatedly enters the fluid and causes coating irregularities. The process is continued once the contact angle regains a value less than 180° and the mechanism of interfacial 'stick and slip' is repeated.

2 WHAT IS AIR ENTRAINMENT?

When a solid surface moves relative to a liquid, and that process takes place in an environment of a third phase, in this case air, eventually a speed of the moving solid surface will be reached at and after which air will be entrained at the three-phase junction, shown in Figure (1).

By air entrainment we mean that air will be drawn along with the continuous surface as an intermediate layer between the coating or contacting fluid of interest, and the surface itself. Because the phenomenon of air entrainment depends to a large extent on the physical properties of the fluid together with the dynamic wetting

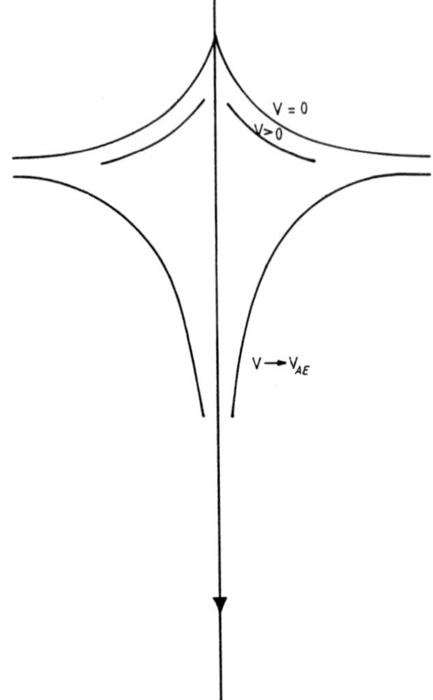

Figure 1 Three-Phase Junction: Moving Solid Surface/Coating Liquid/Air

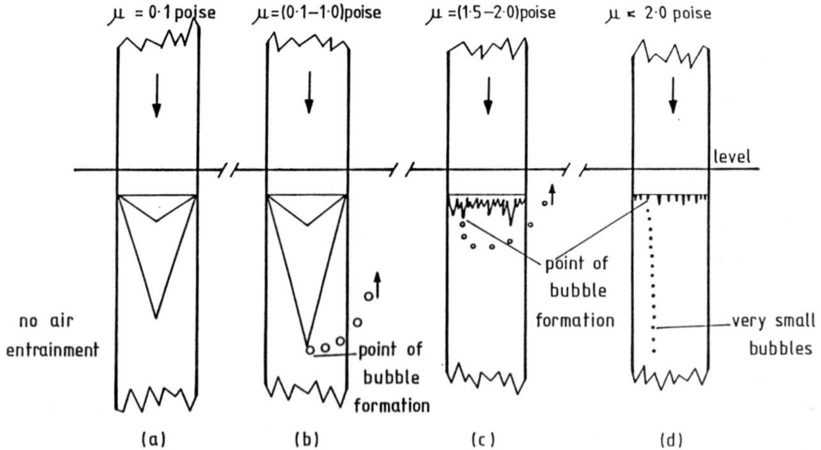

Figure 2 'V' Formation of Air as the Liquid surface become deformed at high speeds of contacting.

properties of the solid surface, air will always finally be entrained at and through the three-phase junction.

Many studies summarised in[1], have shown that above a pre-determined coating speed, the coating fluid layer becomes unstable and discontinuous as the surround medium, usually air, become entrained between the solid surface and the liquid coating. This speed is known as the air entrainment velocity, V_c or V_{ae}.

The air, entrained as an unstable layer between liquid and substrate, subsequently coalesces to form bubbles large enough to detach from the solid surface to which they are temporarily attached. Such bubbles can then be observed as air bubbles rising to the liquid surface. In turn this can lead to the formation of froth and/or foam, which causes further coating problems or re-entrainment preventing full contact between solid surface and contacting fluid. The efficiency of the coating operation decreases and from this point alone air entrainment should be avoided. (It is important to note that whilst it is clear that with no air present, air will not be entrained, the free coating fluid surface will always be depressed at the plunge point, due to the fluid viscous drag on the submerged part of the continuous solid tape. This has been demonstrated convincingly by Robertson[2], with silicone liquids of low vapour pressure. However coating operations in a low pressure environment do not provide realistic processing conditions).

3 WHAT FORM DOES AIR ENTRAINMENT TAKE?

At or beyond the air entrainment velocity, a classical 'V' - shaped intrusion forms between the continuous solid surface and the contacting fluid as shown in Figure (2). The 'V'-shape is actually a plan view of a three-dimensional inverted conical shaped surface. It is independent of the type of equipment used and a common limiting production speed due to will occur for a given fluid. At the apex of a given intrusion, air entrains in the form of bubbles via an intermittent breakage of the wetting line as the dynamic contact line breaks down with the contact angle tending to 180°.

Individual industries have their own terms for such intrusions, but they are normally shaped as shown in Figure (2). The size of the intrusions decreases directly with increase in viscosity range but the 'V' shaped intrusions remain geometrically similar. The shape effectively takes the form of an arrow-head with an included angle lying in the range of 32° to 36°.

The bubbles of air at entrainment may either continue as an attached layer along with the continuous surface, or may detach directly into the surrounding fluid. The smaller, highly stable bubbles are more likely to continue attached to the surface and inhibit contact. Larger bubbles detach and may form foams and emulsions, either of which will be detrimental to any coating or contacting process.

4 MECHANICS OF AIR ENTRAINMENT

Experimental Studies

Studies concerned with the mechanics of air entrainment have been carried out in detail by various workers since 1975.[7,8,11] The studies have examined the plunge point of a moving continuous solid surface into a fluid. The contacting fluid had known properties. This system represented a typical solid/liquid/gas system which met at the three-phase contact line. The unstable movement of this line, due to many primary and secondary causes allows air to be entrained at the point of plunge.

Liquids employed in the studies typically covered a wide range of viscosities i.e. 0.04 - 19.3 poise. Interfacial tension varies, but not so significantly. However a reasonable range of interfacial or surface tensions, i.e. 25 - 65 dynes/cm, were obtained. Both pure liquids and liquid mixtures were employed.

Detailed examination of the point of entry of the high-speed surface was recently achieved by O'Connell,[3] using a 'NAC' High-Speed Video which was capable of shutter times of up to 1/10,000 sec. at 200 or 400 frames per second. Magnification was of the order of 1:30 and achieved using a 105mm telephoto lens plus spacers.

The experiments were carried out in a darkened laboratory using a 150W white light directed through a diffuser fitted with a black covering and an orifice of approximately 1½" diameter.

As shown in Figure (3) this produced excellent illumination of the entry point and demonstrates the contrast between entrained air and the bulk liquid. Further unique photographs may be seen in the original thesis due to O'Connell[3].

Fluid Behaviour at the Plunge Point

At the plunge point, shown in Figure (4), several dynamic processes occur simultaneously. As the continuous surface enters the fluid it draws along with it a boundary layer of air due to viscous drag. Similarly, air is drawn to the point of plunge by the mobile coating liquid, as it also acts as a solid surface with respect to the surrounding air. An excess of air thus occurs at the point of plunge which having its own momentum creates a force which, should the wetting line break down, will cause the instantaneous entry of air along with the continuous surface.

At the same time, the viscous drag caused by the fluid acting below the plunge point on the solid surface, causes a depression of free coating fluid level and a change in the dynamic contact angle towards 180°.

With increasing continuous surface velocity, the air drawn to the point of plunge is expelled in the reverse direction, whilst increasing the force due to its momentum. Simultaneously, the dynamic contact angle increases towards 180° and finally, depending

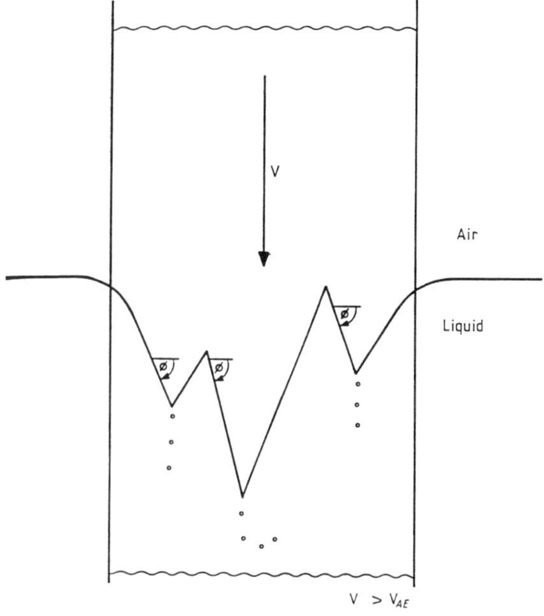

Figure 3 'Saw-Tooth' Wetting Line.

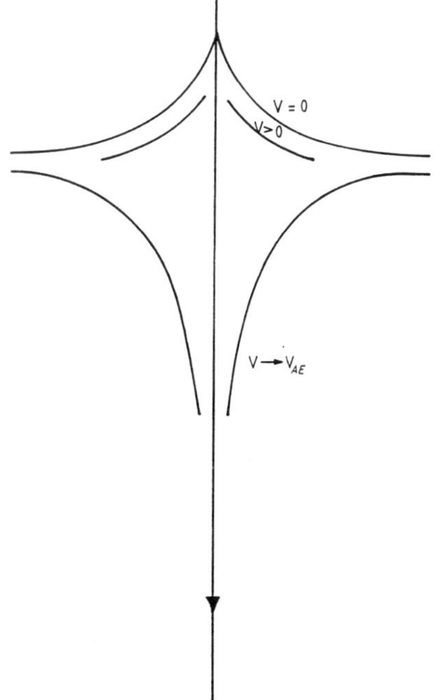

Figure 4 Dynamic Processes at Plunge Point.

on the properties of the coating fluid, air is admitted passed the instantaneously unstable contact line. The surface of the surrounding fluid is locally depressed, due to the submerged viscous drag, and whilst this enhances the onset of air entrainment, the depression is never particularly significant, typically less than 1.0 cm.

The process is unstable and oscillatory, giving the effect of a 'dancing' interface, deriving from 'stick and slip' motion, which alternates between a condition of contact and viscous drag through to non-contact and immediate air entrainment.

Nature of Entrained Air

When the dynamic contact angle tends instantaneously to 180°, air is admitted passed the wetting line as shown in Figure (5). It has been found that the air is admitted in the form of small bubbles, the size of which is dependent on the viscosity of the fluid.

At the point of the 'V'-shaped intrusion below the fluid surface, small quantities of air are admitted, immediately forming bubbles as the most stable form. It has been found that the larger the viscosity of the fluid, the smaller and more stable the bubbles of air which are formed. Such tiny bubbles tend to attach themselves to the surface, thus inhibiting contact between liquid and solid, before slowly coalescing and releasing into surrounding fluid.

Larger bubbles, admitted by less viscous fluids, tend to separate more or less immediately and may form froths or foams at the liquid surface. The process is hydrodynamically controlled so that changes in the nature of the gas phase, unless the gas were highly soluble would not be expected to alter the general nature of the entrainment process.

Prediction of Critical Speeds of Air Entrainment

Having considered the detailed nature of air entrainment, prediction of its onset is of great value in any coating or liquid contacting process. The velocity at this point is known as the air entrainment velocity V_{ae}. At or above this velocity, the surrounding air is unstably entrained into the bulk fluid.

This process of dynamic air entrainment limits the velocity of high speed coating and hence the rate of production of coated material.

Considering the importance of prediction of the air entrainment velocity, relatively few studies have been carried out, but those that have been reported all agree with the general observation that the higher the viscosity of the coating solution, the lower the speed of air entrainment.

Deryagin and Levi[4] appear to be the first workers to have studied the dynamic wetting of flat solid surfaces at speeds of up

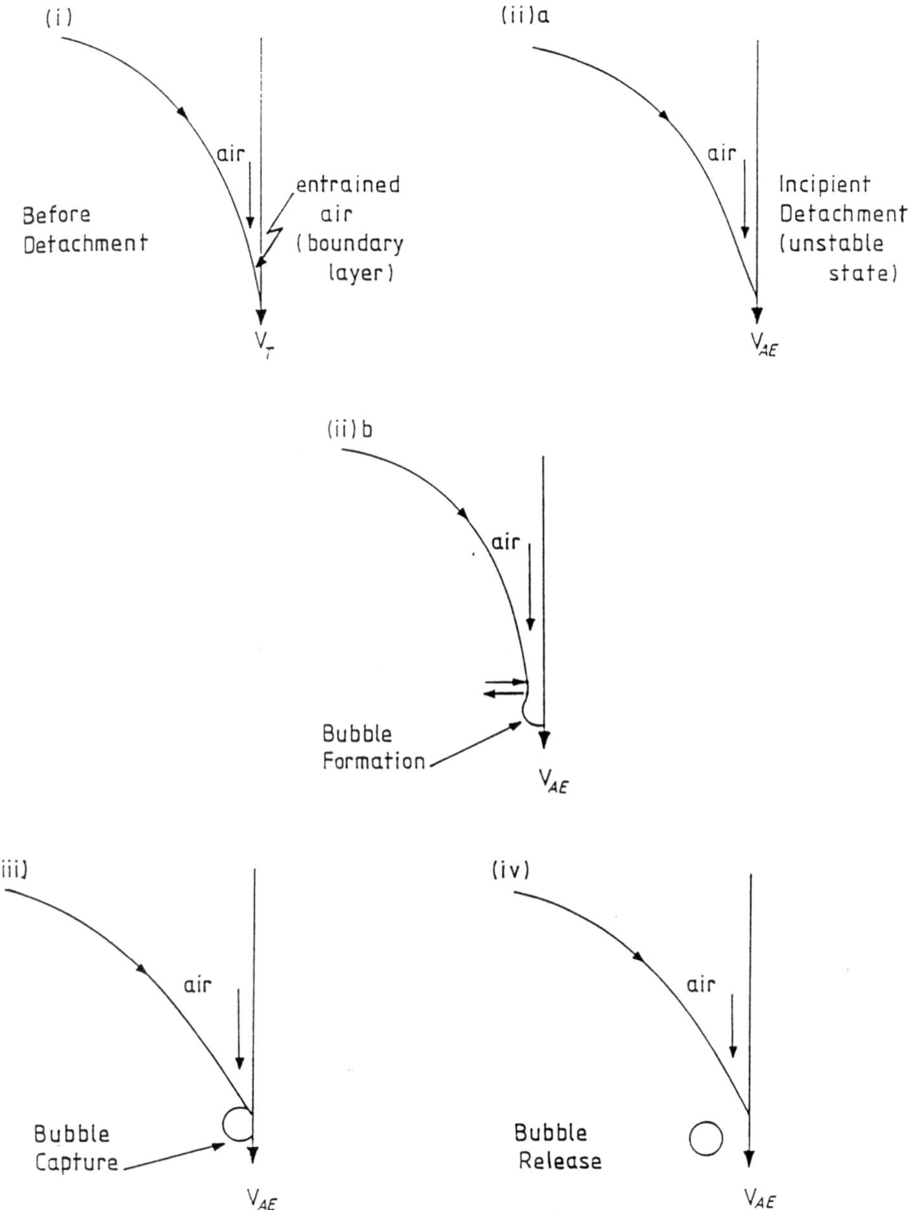

Figure 5 Mechanism of Air being admitted past the wetting line.

to 80 cm/s. Deryagin and Levi's study concerned coating of an emulsion onto a moving surface. As described above as the velocity of the surface increased, the dynamic contact angle increased and eventually tended to 180° at the critical velocity. The first record of a "V - shaped" air film (see Figure (2) was also made between the moving surface and the emulsion. In earlier studies Perry[9] used magnetic recording tape for use as the continuous solid entering the liquid bath. Due to different roughness and wetting characteristics on each face of the magnetic tape values of air entrainment velocity were recorded.

More recently studies have been carried out since 1970 by Burley and co-workers. Kennedy[5,6,14] and Jolly[8,17] conducted experiments into high speed air entrainment employing continuous solid surface with a range of liquids with different physical properties.

In the primary study, Burley and Kennedy[5,6,7,10] showed that air entrainment could be predicted to within 10% of the experimentally determined value with the following equation.

$$V_{ae} = 0.00957 \left[\frac{\sigma^5}{\sigma\,\mu^4}\right]^{(1/6)} + 0.02782 \quad \text{m sec}^{-1} \qquad (1)$$

with the regression coefficient r = 0.99 for liquid properties in the range: $0.04 < \mu < 5$ poise and $25 < \sigma < 64$ dynes/cm.

It was also found that as the velocity of the solid surface increased, a constant value of air entrainment velocity of 0.1 m/s occurred at a viscosity of 4.62 poise. They reported on limited data that for fluids with a higher viscosity the air entrainment velocity became independent of the fluid surface tension at a velocity of approximately 0.1 m/s.

Studies by Gutoff and Kendrick[11] did not determine a lower limit to air entrainment velocity, but confirmed the general form and nature of air entrainment. They developed similar correlations:

$$V_{ae} = 5.11\,\mu^{0.67} \qquad (2)$$

with regression coefficient r = 0.99, whose data agreed with that of Kennedy extremely well.

Burley and Kennedy also reported a function in the form $We = k\,Re^a$ and the best experimental data was shown to be:

$$We = 0.834\,Re^{0.798} \qquad (3)$$

for the following range of Reynold's and Weber Numbers:

$$1.0 < Re < 10^3 \;;\; 1.0 < W_e < 10^2$$

Surface tension was seen to have its greater effect on air entrainment velocity at low viscosity values of the coating liquid. A decrease in surface tension at constant viscosity lowered the air entrainment velocity. However, on increasing the viscosity up to and beyond approximately 4.62 poise the value of air entrainment became unaffected by surface tension.

Practical Correlations for Air Entrainment Prediction

Several studies have produced correlations for the onset of air entrainment, which given their independence and the difficulty of experimental observation, show remarkable agreement.

Burley and Kennedy[5,6,7] for example, used the data from forty separate runs using nine fluids and three tapes.

Their results are represented by the following relationship:

$$V_{ae} = 67.88 \left[\mu \frac{g}{\varnothing\,\sigma}\right]^{-0.672} \qquad (4)$$

which for limited density variation simplifies to:

$$V_{ae} = 1.14 \left[\frac{\sigma}{\mu}\right]^{0.77} \qquad (5)$$

The data may also be represented in an alternative form:

$$V_{ae} = 0.585 \left[\frac{\phi X_b}{\mu}\right]^{0.664} \left[\frac{\sigma}{\phi X_b}\right]^{0.832} \qquad (6)$$

In an earlier study Wilkinson[15] used the data recorded from a rotating pre-wetted cylinder. This study predicted that air entrainment was characterised by a constant capillary number of 1.14. However, re-plotting the limited data shows that:

$$V_{ae} = 1.09 \left[\frac{\sigma}{\mu}\right]^{0.881} \qquad (7)$$

A further in-depth study was carried out by Gutoff and Kendrick[11] whose studies focused on dynamic contact angles developed by a plunging tape entering a pool of liquid. Their dimensional correlation of the form:

$$V_{ae} = 5.11\ \mu^{-0.67} \quad (r = 0.95) \qquad (8)$$

was obtained from the data.

Burley and O'Connell also have recently extended the work of Kennedy and Jolly. Their studies found that the two types of dimensional correlation utilised the fluid properties of viscosity and surface tension and is of the form:

$$V_{ae} = 2.29 \left[\frac{\sigma}{\mu}\right]^{0.637} \qquad (r=0.97) \qquad (9)$$

O'Connell's work also refined the point where viscosity alone controlled the onset of air entrainment to 7.0 cm sec^{-1}. Full details of comparative predicted air entrainment velocities and values of the constants reported for the above equations have been reported previously.[1]

Summary and Conclusions

A graph comparing the prediction of air-entrainment velocities from several studies on a log_{10} basis is given in Figure (6). The graph shows that all the studies provide the same type of curve and there is clear evidence that as the viscosity increases the air entrainment velocity and hence the limit of coatability decreases to a limiting value of both variables. The study recently carried out by O'Connell[3] shows good agreement with the other workers and confirms Burley and Kennedy's [5,6,7] work that originally suggested a constant air entrainment velocity at and beyond a certain coating fluid viscosity. The reported results that above a viscosity of 4.62 poise a constant air entrainment velocity of approximately 10.0 cm sec^{-1} which has now been estimated[3] at 7.0 cm sec^{-1}. Given the nature of the asymptotic curve, however, it is very hard to define an exact figure. Figure (6) shows the curves which demonstrate that at the region of a higher viscosity the air entrainment velocity clearly tends to a constant value at a viscosity of between approximately 4 and 5 poise. A summary of all reported studies is given in Table (1).

Due to the economics of coating and the highly competitive nature of the industry there has been a surge of interest in this work in the last fifteen to twenty years. However, it is unlikely that prediction of the limits of coatability can be further refined as it may be seen that all the studies reported are more or less in accord with differences arising due to secondary effects including pre-wetting, surface roughness, surface charge, surface energy and physico-chemical nature of the surfactants added.

Specific problems may arise due to the particular surface, e.g. wires, filaments, fibres etc., but the general rule will be as predicted. Coating onto complex surfaces, eg fibre matrices, textiles etc. will need further consideration as will the nature of an added surfactant. Further research on these aspects perhaps suggest the need for centres for coating technology so that the process industries who practise this interdisciplinary technology have access to all the available information.

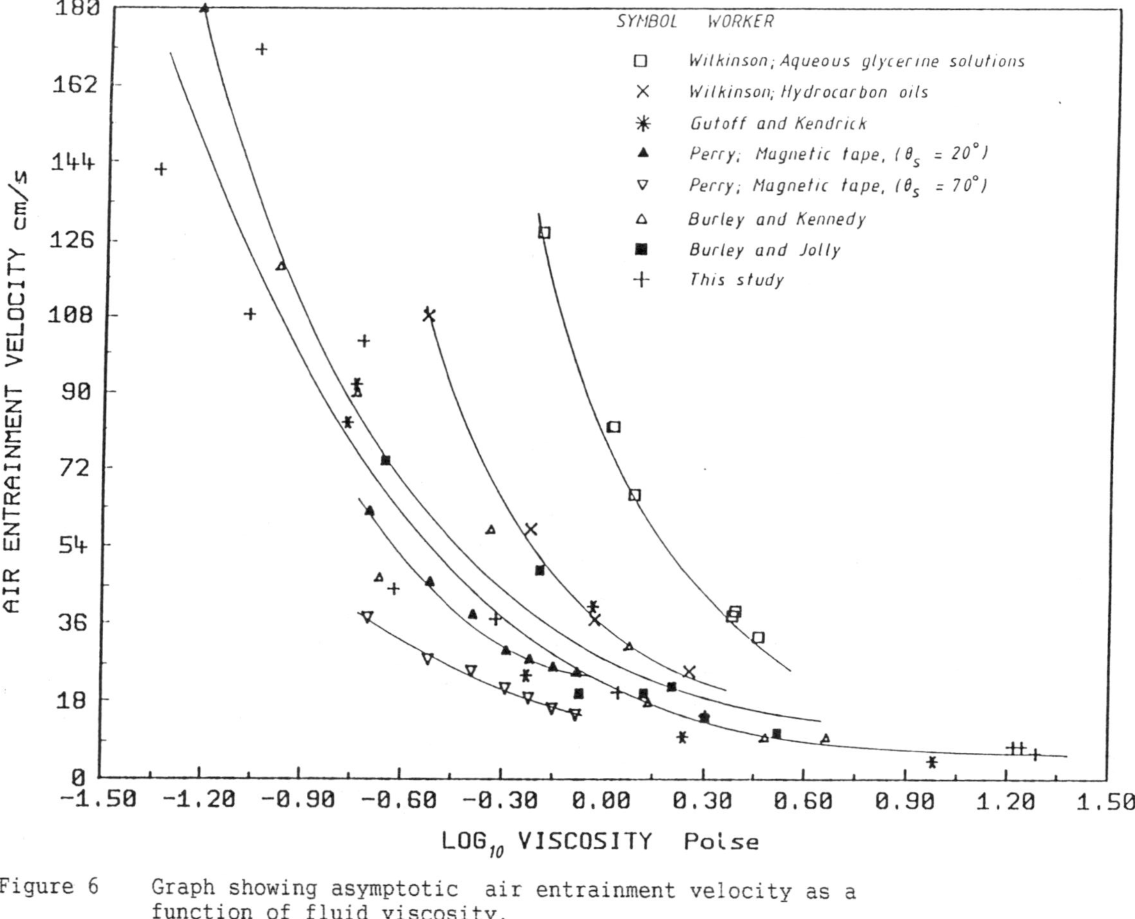

Figure 6 Graph showing asymptotic air entrainment velocity as a function of fluid viscosity.

Acknowledgements

The support of Industry and the SERC, Process Engineering committee is gratefully recorded. Industries who have assisted since 1975 include ICI (Mond), Union Carbide, ICI (Wilton), Kodak (UK), Eastman-Kodak, (USA), Polaroid (USA), and British Cellophane. The author also wishes to record his gratitude to Dr Kennedy, Dr Jolly and Dr O'Connell for their significant contribution of new knowledge in the resolution of the problem of high-speed coating.

5. REFERENCES

1. Burley, R, 1991, JoCCA Trans. 5, 191-202 "Air Entrainment and the Limits of Coatability".

2. Robertson, W G C, O'Shaughnessy, D P, and Molloy, N A, 1973 Chem Eng Sic 28 1635-1637.

3. O'Connell, A 1989 PhD Thesis, Heriot-Watt University, Dept Chemical and Process Engineering: 'Observations of Air Entrainment and the Limits of Coatability'.

4. Deryagin, B V and Levi, S M 1964 Film coating Theory, Focal Press, London.

5. Kennedy, B S and Burley, R ; J Colloid Interface Sc. 62 No 1 (1977).

6. Burley, R and Kennedy, B S ; Brit. Polymer J 8, 140, (1976).

7. Burley, R and Kennedy, B S ; Wetting, Spreading and Adhesion, P327 ed. Padday, Academic Press (1978).

8. Burley, R and Jolly, R P S, (1984) Chem Eng Sci 39 9, 1357-1372 also Burley, R and Jolly, R P S, AIChE Journal, 29, 1, 173, (1983).8.

9. Perry, R T, ; PhD. Thesis, Univ. of Minnesota, Mineapolis, USA (1967).

10. Kennedy, B S, PhD. Thesis, Heriot-Watt University, Edinburgh, (1975).

11. Gutoff, E B and Kendrick, G E, AIChE. Journal, 28 3, (1982).

12. O'Connell, A & Burley R. Observations of Air Entrainment AIChE Conference Orlando, March 1990. Mechanics of Thin Film Coating.

13. Blake, T D, Wetting Kinetics - How do wetting Lines Move? 1988 AIChE. International Symposium on the Mechanics of Thin-Film coating, New Orleans (1988).

14. Burley, R and Kennedy, B S, ; Chem. Engineering Sci., 31, pp 901-911, (1976).

15. Wilkinson, W L ; Chem. Engineering Sci., 30 1227, (1975).

16. Miyamoto, K Industrial Coating Research, 1 No 1, 89-96 Sept 1991 Coating Research Association Kyoto University, Japan 606.

17. Jolly, R P S PhD Thesis, Department of Chemical and Process Engineering, Heriot-Watt University, 1982.

Nomenclature

g acceleration due to gravity
V_{ae} velocity of air entrainment
X_d dispacement distance
ϕ fluid density
σ surface or interfacial tension
μ fluid viscosity
α contact angle
B wetting line angle
U linear tape velocity

Re Reynolds Number: Ratio Momemtum/Viscous Forces
Ca Capillary Number: Ratio Viscous to Interfacial Forces
We Weber Number: Ratio Momentum to Interfacial Forces

Drying

The Basis of Dryer Design

Jeremy M. Double and Hadj Benkreira
DEPARTMENT OF CHEMICAL ENGINEERING, UNIVERSITY OF BRADFORD, BRADFORD BD7 1DP, UK

1 INTRODUCTION

This paper discusses the design of dryers used in the drying of coatings. Figure 1 shows the general arrangement of the type of dryer considered in this paper. This is a simple dryer with recycle of part of the drying air.

The material and energy balances around the dryer are discussed first, followed by one possible procedure for solving these equations.

The specification of dryer length is discussed next. This involves consideration of mass and heat transfer in the dryer. A correlation for calculating heat transfer coefficient is quoted, which shows the influence of gas velocity in the dryer.

This is followed by a brief discussion of the trade-offs involved in the dryer specification, particularly considering the influence of recycle ratio.

2 BASIC BALANCE EQUATIONS

To assist in the calculations, gas stream variables are quoted on a dry air basis, and variables concerned with the coating are quoted on a dry solids basis.

Air streams are defined by their flowrate (G), humidity (H) and temperature (t_g). The specific enthalpy of the air stream (i) is specified relative to a base of dry air at 0°C. It includes both the sensible heat of the air and associated solvent vapour and the latent heat of the vapour. It is a function of both the temperature and the humidity, and can be found by using the psychrometric chart of the relevant solvent-air system.

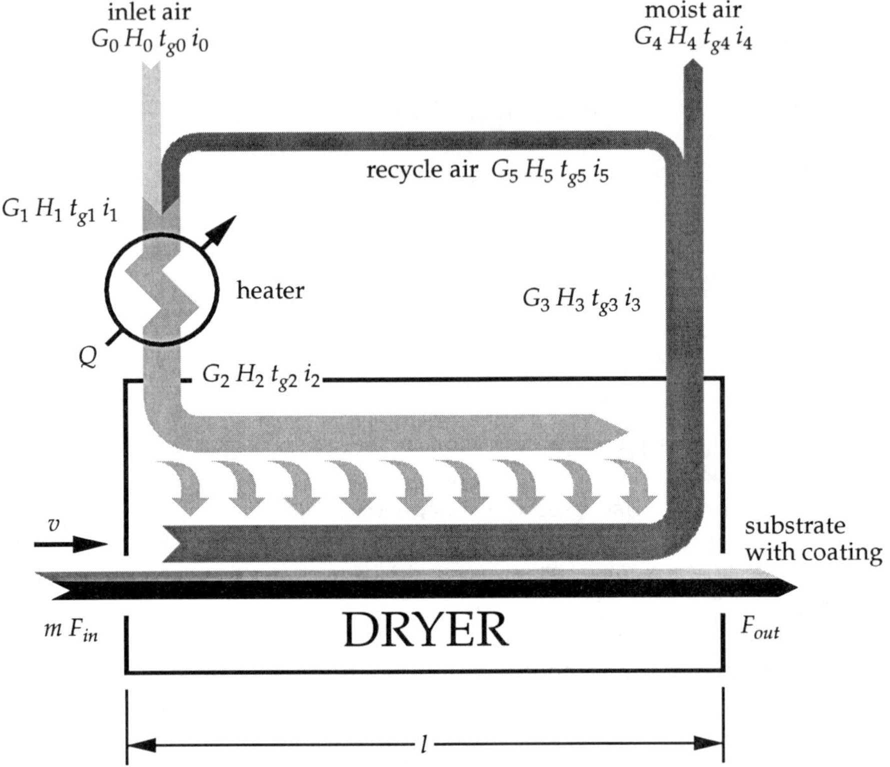

Figure 1 Simplified diagram of a dryer with recycle

The coating is defined in terms of its application density per unit area (m) and its moisture content (F). The width of the coated strip (b) and the velocity which it passes through the dryer (v) must also be specified.

The following relationships link gas flow rates in the system:

$$G_1 = G_2 = G_3 \quad [1]$$

$$G_0 = G_4 \quad [2]$$

The following relationships link humidities in the system:

$$H_1 = H_2 \quad [3]$$

$$H_3 = H_4 = H_5 \quad [4]$$

An overall material balance on the solvent (also using equation [2]) yields the following relationship:

$$N = G_0 \left(H_4 - H_0\right) = b\, m\, v \left(F_{in} - F_{out}\right) \qquad [5]$$

Ignoring the small amount of heat required to heat the substrate and coating to process temperature gives the following overall energy balance:

$$Q = G_0\left(i_4 - i_0\right) \qquad [6]$$

Material and energy balances over the junction of the inlet and recycle streams:

$$G_1 = G_0 + G_5 \qquad [7]$$

$$G_1\, H_1 = G_0\, H_0 + G_5\, H_5 \qquad [8]$$

$$G_1\, i_1 = G_0\, i_0 + G_5\, i_5 \qquad [9]$$

If the small amount of heat required to heat the substrate and coating to process temperature is again ignored, energy balances and equation [1] are combined to give:

$$i_2 = i_3 = i_4 = i_5 \qquad [10]$$

3 CALCULATION PROCEDURE FOR BALANCE EQUATIONS

Usually, the dryer will be specified in terms of b, v, m, F_{in} and F_{out}. The duty of the dryer, N, is thus easily calculated using equation [5].

Typical values of H_0 and t_{g0} can be estimated from regular measurements in the dryer's location. In this paper, it is assumed that the designer chooses the values of H_4, t_{g4} and the recycle ratio, R, defined as:

$$R = \frac{G_5}{G_4} \qquad [11]$$

H_4 and t_{g4} must be chosen so that H_4 is below the saturated humidity at t_{g4}, so that there is a driving force for evaporating solvent in the dryer.

The calculation procedure is then as follows:

- G_0 is calculated using equation [5]. G_4 is equal to G_0 (equation [2]).
- Knowing H_0 and t_{g0}, i_0 can be found using a psychrometric chart.
- Similarly, i_4 can be found from H_4 and t_{g4}. H_3 and H_5 are equal to H_4 (equation [4]) and i_2, i_3 and i_5 are equal to i_4 (equation [10]). It can be inferred that t_{g3} and t_{g5} are equal to t_{g4}.
- Q can then be found using equation [6].
- G_5 is calculated from R using equation [11].
- G_1 can then be calculated using equation [7]. G_2 and G_3 are equal to G_1 (equation [1]).
- H_1 is calculated using equation [8]. H_2 is equal to H_1 (equation [3]).
- i_1 can then be calculated from equation [9].
- t_{g1} can be found from H_1 and i_1 using a psychrometric chart.
- Similarly, t_{g2} can be found from H_2 and i_2.
- Temperature t_{g2} should be checked to make sure that it is not too high for the material being dried.

4 CALCULATION OF DRYER LENGTH

The calculations discussed above have only used the material and energy balances, without consideration of the size of the dryer required. In order to determine the length of dryer required, it is necessary to consider the mass and heat transfer that take place during the drying process.

Figure 2 shows the processes that occur at a microscopic scale in the dryer, while the coating is still wet at its surface.

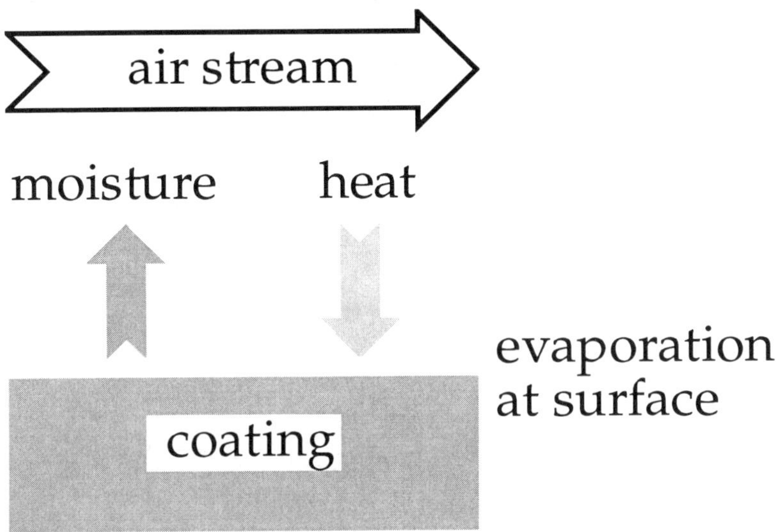

Figure 2 Heat and mass transfer

Solvent evaporates at the surface of the coating, and diffuses into the air stream flowing across the surface. In order for this to happen, the latent heat of vapourisation of the solvent must be provided, so there is a transfer of heat from the air stream to the surface of the coating.

5 HEAT AND MASS TRANSFER THEORY

In either heat or mass transfer, the rate of transfer per unit area is equal to a driving force multiplied by a transfer coefficient. Referring to figure 3, the mass transfer is described by:

$$n = k_g\,(H_i - H) \qquad [12]$$

The heat transfer is described by:

$$q = h_g\left(t_g - t_{g,i}\right) \qquad [13]$$

Assuming that all of the latent heat required for evaporation of the solvent is provided by heat transfer from the air stream, the following equation may be inferred:

$$n = k_g\,(H_i - H) = \frac{q}{\lambda} = \frac{h_g}{\lambda}\left(t_g - t_{g,i}\right) \qquad [14]$$

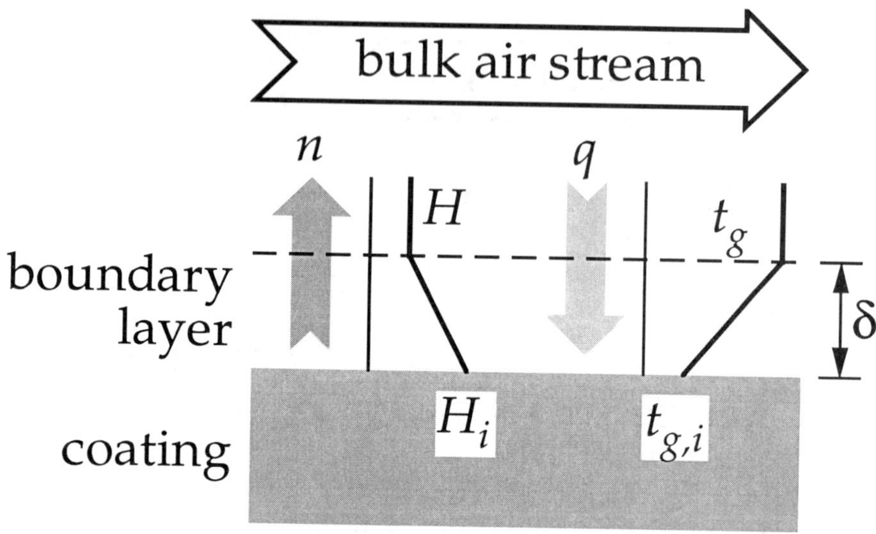

Figure 3 Boundary layer in mass and heat transfer

Conventionally, the humidity and temperature profiles are assumed to be as shown in figure 3: a straight line profile from the surface of the coating to the outer edge of the boundary layer, and a constant value of temperature and humidity in the bulk fluid. The fluid is assumed to be in laminar flow in the boundary layer: heat is transferred by conduction through the layer and mass transfer is by diffusion.

The transfer coefficients can therefore be considered to be:

$$k_g = \frac{D}{\delta} \quad [15]$$

$$h_g = \frac{K}{\delta} \quad [16]$$

where D is a diffusion coefficient for solvent vapour in air and K is the thermal conductivity of the moist air. It can be seen that each of these coefficients is dependent on the thickness of the boundary layer, δ. The boundary layer thickness is determined by the fluid flow regime in the system (for a given geometry, this will be measured by the Reynolds number, Re). A high air velocity (and hence a high value of Re) will reduce the boundary layer thickness, δ, and hence increase the values of h_g and k_g. Thus, although the values of h_g and k_g depend strongly on the air velocity (and hence Re), the ratio of the two coefficients is a constant.

Considering the conditions at the coating-air interface, the air at the interface will be in equilibrium with the solvent in the coating. The humidity H_i will be the saturated humidity at temperature $t_{g,i}$. Supposing the value of h_g is known, equation [14] cannot be solved for the rate of drying, n, because $t_{g,i}$ is not known. If k_g is known, the same logic applies, because $t_{g,i}$ is not known so H_i will not be known.

However, if the flow of the air stream is large compared with the amount of moisture evaporated, the temperature at the interface, $t_{g,i}$, will approximate to the wet bulb temperature of the air stream. This can be found using a psychrometric chart given the temperature and humidity of the air stream. Thus, if h_g can be estimated, then the rate of drying, n, can be found using equation [14]. From this it is easy to calculate the surface area required in the drier, and hence its length.

6 ESTIMATION OF HEAT TRANSFER COEFFICIENT

A number of correlations have been published for estimating the heat transfer coefficient in drying. A simple relationship, developed by Shepherd [1] was as follows:

$$h_g = c\,G^{0.8} \qquad [17]$$

where c is a constant which depends on the geometry. More recent correlations are in the form of dimensionless equations.

Figure 4 shows the geometry of a slot through which air is blown onto the surface of the coating in the drier. For this arrangement [2]:

$$Nu = 0.135\,Re^{0.697}\left(\frac{s}{w}\right)^{-0.351} \qquad [18]$$

where:

$$Nu = \text{Nusselt number} = \frac{h_g\,w}{K} \qquad [19]$$

$$Re = \text{Reynolds number} = \frac{\rho\,u\,w}{\mu} \qquad [20]$$

For this correlation to give good predictions of the length of the dryer, the air temperature in equation [14] must be t_{g2}.

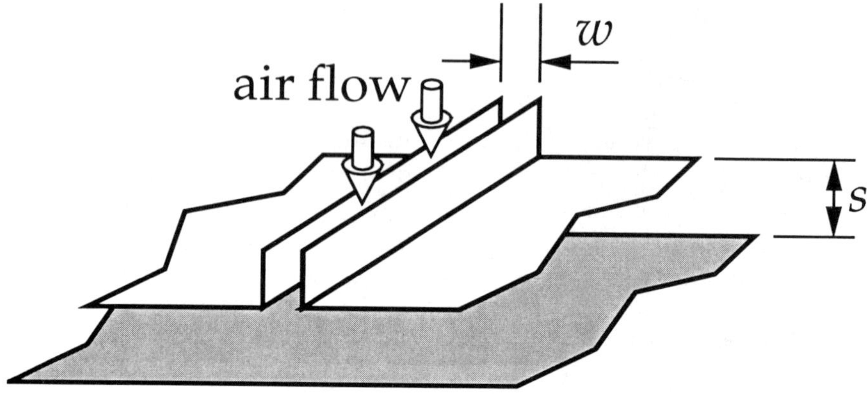

Figure 4 **Dimensions of an air slot within a dryer for equation [18]**

7 FALLING RATE DRYING

The analysis presented above has assumed that the surface of the coating is wet. At some point, the surface of the coating ceases to be wet, and moisture must diffuse through the solid before it reaches the surface. As the coating dries out, the distance through which the solvent must diffuse increases, thus increasing the resistance to mass transfer. The rate of drying will therefore fall as the moisture content decreases. However, in the case of a thin coating, this stage of drying is likely to be unimportant compared with the constant rate period described above.

This stage of drying can be dealt with using empirical data, or by using sophisticated models of the heat and mass transfer that take place within the solid. This is outside the scope of this paper. See, for instance, Keey [3].

8 CHOICE OF RECYCLE RATIO

From the above analysis, it can be inferred that a high air velocity is required to give a good heat transfer coefficient, and hence a manageable length of dryer. However, a high air velocity implies a high power consumption in the fan supplying air to the dryer. There is thus a trade off between capital cost (length of dryer) and operating cost (power usage).

Interacting with this calculation is the consideration of recycle ratio. If a low recycle ratio is used, then to get a sufficient air velocity, a high inlet air flow rate will be required, and hence a lot of energy must be supplied which will be lost in the exit air stream. However, stream 1 will be drier than if a higher recycle ratio is used, because moisture will not be carried into it from the recycle stream. This will increase the driving force for mass transfer, reducing slightly the air velocity required for a given rate of drying.

In contrast, a high recycle ratio will mean a smaller difference between inlet and outlet conditions in the dryer, improving controllability. Only a small inlet and exit air stream will be required, reducing the waste of heat from the dryer. However, the driving force for mass transfer may be reduced.

If the solvent is to be recovered from the exit air stream, a high recycle ratio is advantageous, since the separation process will be handling a smaller quantity of air at a higher solvent concentration. This will lead to improved collection efficiency and lower capital and operating costs.

9 CONCLUSION

It can be seen that the design parameters of a dryer are related in a complex manner, with both balance equations and rate equations playing an important part in the final results. It is important to make sure that the balance equations are taken into account in the design of dryers, so that the correct driving forces for the heat and mass transfer equations are used.

REFERENCES

1. C B Shepherd, C Hadlock and R C Brewer, Ind. Eng. Chem., 1938, 30, p 506.
2. R B Keey, "Introduction to industrial drying operations", Pergamon, Oxford, 1978.
3. R B Keey, "Drying: principles and practice" Pergamon, Oxford, 1972.

NOMENCLATURE

Note: numeric subscripts refer to the properties of the numbered streams in figure 1. The subscripts *in* and *out* refer to properties of the coating at the inlet to and exit from the dryer. The subscript *i* refers to conditions at the interface.

b	Breadth of substrate and coating, m
D	Diffusion coefficient for solvent vapour in air, kg/m s
F	Free moisture content of coating, kg solvent/kg dry solid
G	Flowrate of air, expressed as flowrate of completely dry air, kg/s
h_g	Heat transfer coefficient, W/m^2 K
H	Humidity of air, kg solvent vapour/kg dry air
i	Specific enthalpy of air, kJ/kg dry air
k_g	Mass transfer coefficient, kg/m^2 s
K	Thermal conductivity of moist air, W/m °C
l	Length of dryer, m
m	Application density of coating, kg dry solids/m^2
n	Mass flux of solvent, kg/m^2 s
N	Quantity of moisture evaporated, kg/s
q	Heat flux, J/m^2 s
Q	Heater duty, kJ/s (≡ kW)
R	Recycle ratio (= flowrate of recycle stream/flowrate of exit stream)
Re	Reynolds number (as defined in text).
t_g	Temperature of air stream, °C
v	Velocity of substrate and coating in the dryer, m/s
δ	Thickness of boundary layer, m
λ	Latent heat of vapourisation of solvent, J/kg

APPENDIX A – EXAMPLE DRYER CALCULATION

Input data

Humidity of inlet air	0.006	
Temperature of inlet air	15	°C
RH of outlet air	90%	
Temperature of outlet air	65	°C
Recycle ratio	7	
Mass of coating/unit area	0.02	kg dry solids/m^2
Solvent content at inlet	1	kg solvent/kg dry solids
Solvent content at outlet	0	kg solvent/kg dry solids
Width of strip being dried	1.8	m
Velocity of strip in dryer	2.5	m/s
Heat transfer coefficient	149	W/m^2 K

Results

Humidity of outlet air	0.178	
Duty of dryer	0.09	kg/s
Heating duty	263	kW
Useful heat	86%	
Length of dryer	18.4	m

Material and energy balances

Stream	0 inlet	1	2	3	4 exit	5 recycle
Air flow, dry basis, kg/s	0.524	4.193	4.193	4.193	0.524	3.669
Humidity, kg/kg dry air	0.006	0.157	0.157	0.178	0.178	0.178
Temperature, °C	15	60	108	65	65	65
Enthalpy, kJ/kg dry air	31.2	469.6	532.2	532.2	532.2	532.2

Heat transfer calculations

Temperature	108	°C
Wet bulb temperature	63	°C
Heat to be transferred	225090	W
Area	33.0	m^2
length of dryer	18.4	m

New Computer Techniques for Optimisation of Drying of Coatings

John C. Ashworth
ADVANCED SOLIDS PROCESSING CONCEPTS (ASPCETS), 'THE KNOLL', OLDWICH LANE EAST, FEN END, NEAR KENILWORTH, WARWICKSHIRE CV8 1NR, UK

ABSTRACT

Computer software suitable for use on the IBM family of microcomputers and compatibles running under DOS has been developed for modelling of processing of continuous webs, called CONTIWEB. Many features are incorporated into the program to make it as user-friendly as possible and suitable for a wide range of applications. For example, a modular approach is adopted to the modelling so whole processing lines can be assembled by use of building-blocks for the unwind reel, coater, infrared radiation panels, jet drying, hot cylinders, web cooling, etc. Extensive databases are also incorporated to allow the modelling of water based and solvent based processes.

The potential value of this software to design and production personnel in the converting industries is illustrated by several examples:

(a) Interaction Between Coating and Substrate - if the operating conditions are badly chosen, unexpected interactions can occur. This is demonstrated by an example using a two-sided nozzle configuration to process a 8 gsm coating on a 70 gsm paper substrate. Up to the halfway point, the substrate gains moisture leading to poor line efficiency for the final 30% of the dryer.

(b) Product Quality Interactions - constraints due to thermal degradation are examined by contour plots of product temperature with respect to air temperature and airstream humidity to help identify the most suitable operating conditions for the process.

(c) Upgrading Linespeed - the scope to enhance productivity is studied by examination of the influence of nozzle velocity, airstream temperature and airstream humidity upon dryer performance. It is shown that plant revamping could enable speed increases of up to 100%.

(d) Minimisation of Energy Costs - optimisation of the operating conditions at a given production rate can lead to substantial reductions in fuel costs and electrical demands for the fans. If the design capacity of the dryer is properly exploited, then savings up to £40,000 per year for each tonne per hour moisture evaporative capacity may be possible.

(e) Design of New Process Line - to find the most suitable plant for a particular application means that a wide range of feasible designs may need to be evaluated. This task can be greatly simplified by use of constraint diagrams which in the authors experience typically leads to capital cost savings of 10% and energy cost savings of 15% over the lifetime of the plant.

INTRODUCTION

Traditionally the design of dryers has been empirical because of the difficulty of characterising solid materials. Over the past 15 years there has been a dramatic increase in drying research and development worldwide. The fundamentals are now sufficiently well understood that an effective methodology (ref. 1) based upon a judicious blend of theory and experiment, now exists for the solution of most drying problems. The essential keys to this scientific method are determination of the processing characteristics of the material and the performance characteristics of the drying plant. It is important to separate these two factors, and it was probably lack of realisation of this fact that stalled the development of effective dryer analysis procedures until recently.

The author's Company has been at the forefront of these developments in drying science and pioneered new techniques and analysis procedures. This includes a sophisticated wind-tunnel/balance system which is used extensively for industrial drying investigations. This facility is instrumented to a high standard and includes video monitoring, macro- and micro-scale, of the material while under test.

This expertise for understanding the performance of process plant has now been encapsulated within process software modelling software suitable for use on desktop microcomputers (ref. 2). The software package called CONTIWEB, suitable for simulation of continuous web processing, has been specially developed to assist technologists in the paper and converting industries. This paper discusses the use of the modelling software to solve common problems associated with questions of productivity, efficiency and product quality.

FEATURES OF CONTIWEB

These include:

* Powerful user friendly interface such that each function can be accessed by single key commands.
* Sophisticated Input Datafile facility for accurate definition of processing application.
* Modular approach to modelling so whole processing line can be assembled by use of building-blocks for each process step, up to maximum of 100 zones.
* Extensive HELP displays available on-line.
* On-line Databases for moisture properties, material properties, gas properties, energy/coolant sources and nozzle heat transfer correlations.
* Capability to examine processing of water-wetted and solvent-based applications including multicomponent moisture problems (maximum of ten moisture components present).
* Flexibility to enable simulation of processing of wide range of products including applications involving manufacture of multilayered materials (maximum of ten layers where each substrate and coating counts as a layer).
* On-line modelling facilities to manipulate operating parameters on a zone by zone basis, optimisation of operating costs, and graphical displays of modelling results.

MENU STRUCTURE

Once the program loading is completed, the MAIN MENU is displayed on the screen:

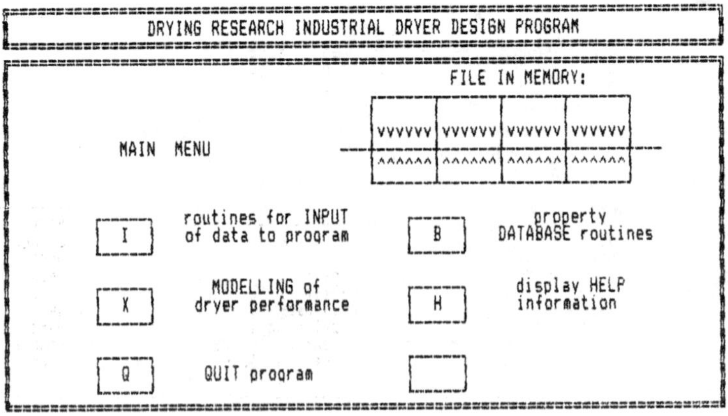

```
DRYING RESEARCH INDUSTRIAL DRYER DESIGN PROGRAM

                                    FILE IN MEMORY:
                              vvvvvv|vvvvvv|vvvvvv|vvvvvv
      MAIN  MENU          ----------------------------------
                              ^^^^^^|^^^^^^|^^^^^^|^^^^^^

  [ I ]  routines for INPUT          [ B ]   property
         of data to program                  DATABASE routines

  [ X ]  MODELLING of                [ H ]   display HELP
         dryer performance                   information

  [ Q ]  QUIT program                [   ]
```

The options displayed on the Menu as shown by the mock keytops are accessed by entry of the corresponding single key instruction.

INPUT DATAFILE PROCEDURES

The Input Datafile is arranged with 10 parameters per page, up to a maximum of 600 pages. The entry procedure incorporates a powerful screen editor to assist manipulation of the parameters to define the processing application. Some examples of the typical screen displays used for creating the Input Datafile are displayed below:

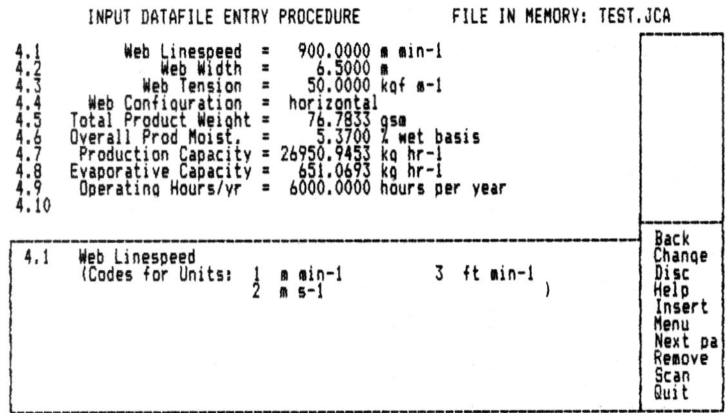

Page 4 of the Datafile is displayed above, and the cursor is positioned for entry of parameter number 4.6. Note the screen is divided into windows. The names, numerical values, and units are displayed at the top. In the bottom half is set out the definition of the present parameter and the choice of units. Down the side the ten editing commands are listed.

```
INPUT DATAFILE ENTRY PROCEDURE          FILE IN MEMORY: TEST.JCA

11.1   Type Process Zone 1 =  unwind reel
11.2   Type Process Zone 2 =  single-sided coating
11.3   Type Process Zone 3 =  convection drying zone
11.4   Type Process Zone 4 =  wind-up reel
11.5
11.6
11.7
11.8
11.9
11.10

11.1   Type of Process Zone no. 1                                  Back
       (Code for  Types:                                           Change
        1  unwind reel        2  wind-up reel        3  s.s. coating    Disc
        4  d.s. coating       5  impregnation        6  open section    Help
        7  IR panel           8  lamination nip      9  compress roll   Insert
       10  convect drying    11  conv/hotcyl cond   12  conv/IR drying  Menu
       13  conv/IR/heat rec  14  conv/HF heating    15  conv/induct heat Next pa
       16  humi-air condit   17  steam condition    18  conv cooling    Remove
       19  chill cyl cool    20  air-turn            )                  Scan
                                                                   Quit
```

This Datafile page illustrates the way the process line is specified as a number of modules in series. Twenty different types of zone are available.

```
         INPUT DATAFILE ENTRY PROCEDURE         FILE IN MEMORY: TEST.JCA
                                                                   ----------
253.1  Air Velocity(above) =      50.0000 m s-1                    |        |
253.2  Air Velocity(below) =      50.0000 m s-1                    |        |
253.3  Static Press(above) =      -0.0100 ins water                |        |
253.4  Static Press(below) =      -0.0100 ins water                |        |
253.5     Air Temp (above) =     285.0000 C                        |        |
253.6     Air Temp (below) =     285.0000 C                        |        |
253.7    Air Humid (above) =       0.2500 kg kg-1 or lb lb-1 dry basis|     |
253.8    Air Humid (below) =       0.2500 kg kg-1 or lb lb-1 dry basis|     |
253.9   Air Solvent(above) =       0.     kg kg-1 or lb lb-1 dry basis|     |
253.10  Air Solvent(below) =       0.     kg kg-1 or lb lb-1 dry basis|     |
      Type of Process Zone : Convection Drying Zone                ----------
---------------------------------------------------------------------| Back
|253.1   Air Velocity/Nozzle Box Pressure (above) for Zone 3         | Change
|        (Note the air velocity value must take proper account of   | Disc
|         the static pressure developed at the nozzle in relation   | Help
|         to the cushion pressure.                                  | Insert
|         Code for Units:  1  m s-1           4  ins water          | Menu
|                          2  ft min-1        5  mms water          | Next pa
|                          3  ft s-1          6  Pa           )     | Remove
|                                                                   | Scan
|                                                                   | Quit
---------------------------------------------------------------------------
```

This Datafile page sets out the airstream conditions for zone 3, a convection drying section. Note in this instance nozzles are installed both above and below the web to provide drying on two sides.

DATABASE PROCEDURES

Databases are incorporated into the program to provide the thermodynamic and thermophysical properties of the materials, moisture, and gases being handled in the process. The appropriate property correlations are automatically linked together with the Input Datafile prior to modelling. Facilities are available to edit and display the Database files:

```
===========================================================================
|               MOISTURE PROPERTIES DATABASE (PAGE 5)                     |
===========================================================================
          IDEAL GAS HEAT CAPACITY:              FILENAME: methanol
          1. Number Correlations=     1
 Corr1:   2. Min Limits-Temp=     100.0000    Value=       3.8191E+04
          3. Max Limits-Temp=    1500.0000    Value=       1.0511E+05
          4. Eqn Type =   103 Coeffs: A=    3.8188E+04  B=  1.0424E+05
             Coeffs: C=    2.1867E+03 D=    1.1628E+00  E=  0.0000E+00
 Corr2:   5. Min Limits-Temp=                 Value=
          6. Max Limits-Temp=                 Value=
          7. Eqn Type =        Coeffs: A=                B=
             Coeffs: C=                D=                E=
          SECOND VIRIAL COEFFICIENT:
          1. Number Correlations=     1
 Corr1:   2. Min Limits-Temp=     256.3000    Value=      -9.1876E+00
          3. Max Limits-Temp=     800.0000    Value=      -7.1218E-02
          4. Eqn Type =   104 Coeffs: A=   -6.4000E-01  B=  6.2000E+02
             Coeffs: C=   -1.0710E+08 D=    7.7160E+20  E= -2.1974E+23
 Corr2:   5. Min Limits-Temp=                 Value=
          6. Max Limits-Temp=                 Value=
          7. Eqn Type =        Coeffs: A=                B=
             Coeffs: C=                D=                E=
  Back  Change  File create  Help  Load file  Next page  Page back  Test  Quit
```

Page 5 of the Moisture Properties Database file is displayed above for methanol. This sets out the constants for the correlation equations for the vapour heat capacity and vapour density. Further pages contain correlations and data for vapour pressure, heat of vaporisation, thermal conductivity, flammibility limits, flash point, etc.

```
MATERIAL PROPERTIES DATABASE (PAGE 6)

                                                  FILENAME: paper
NORMALISED DRYING RATE CURVE FUNCTIONS:
 1. NUMBER OF RATE CURVES =    1
DRYING RATE CURVE 1
 2. Moisture Component = water                  3. Temperature/C=      100.0000
 4. Maximum Moisture Content/%db =    60.0000
 5. Drying Data:    0%mc= 0.        5%mc= 0.3140     10%mc= 0.4260     15%mc= 0.5100
    20%mc= 0.5790  25%mc= 0.6340  30%mc=0.6690     35%mc= 0.7030     40%mc= 0.7250
    45%mc= 0.7530  50%mc= 0.7970  55%mc=0.8420     60%mc= 0.8610     65%mc= 0.8720
    70%mc= 0.8860  75%mc= 0.8990  80%mc=0.9110     85%mc= 0.9300     90%mc= 0.9480
    95%mc= 0.9730  100%mc=1.0000
DRYING RATE CURVE 2
 6. Moisture Component =                        7. Temperature/C=
 8. Maximum Moisture Content/%db =
 9. Drying Data:    0%mc=           5%mc=            10%mc=            15%mc=
    20%mc=         25%mc=          30%mc=            35%mc=            40%mc=
    45%mc=         50%mc=          55%mc=            60%mc=            65%mc=
    70%mc=         75%mc=          80%mc=            85%mc=            90%mc=
    95%mc=         100%mc=

 Back  Change  File create  Help  Load file  Next page  Page back  Test  Quit
```

Page 6 of the Material Properties Database file is displayed above for paper. This shows the drying rate curve function of moisture content. Other pages set out data for the heat capacity, thermal conductivity, infrared radiation adsorption, equilibrium moisture content isotherm, loss factor, density, etc.

```
ENERGY/COOLANT SOURCE PROPERTIES DATABASE (PAGE 2)

                                      FILENAME: Commercial Propane (LPG)
                     TYPE 4: FUEL
           DENSITY AT NTP/kg m-3 :      1.8690
     SPECIFIC HEAT CAPACITY/Jkg-1K-1 :   1534.6198
       GROSS CALORIFIC VALUE/MJ kg-1 :     50.2300
         NET CALORIFIC VALUE/MJ kg-1 :     46.2400
   STOICHIOMETRIC AIR/kg per kg fuel :     15.6000
      COMBUSTION CO2/kg per kg fuel :      3.0140
      COMBUSTION H2O/kg per kg fuel :      1.6220

                TYPE 5: THERMAL FLUID
     SPECIFIC HEAT CAPACITY/Jkg-1K-1 :
                       TEMPERATURE/C :
            GENERATION EFFICIENCY/% :

             TYPE 6: REFRIGERANT FLUID
           EVAPORATING TEMPERATURE/C :
    LATENT HEAT VAPORISATION/kJ kg-1 :
            GENERATION EFFICIENCY/% :

 Back  Change  File create  Help  Load file  Next page  Page back  Quit
```

Page 2 of the Database file is displayed above for the Energy/ Coolant source of LPG or commercial propane. This presents the fuel combustion properties. Other types of sources such as electricity, steam, air, thermal fluids, & refrigerant fluids can be defined within the Database.

MODELLING OF PROCESS LINE

The modelling routines calculate the detailed profiles of material and gas moisture, temperature and enthalpy on a step by step basis along the process line. This requires the solution of the set of differential equations which summarise the heat balance, mass balance, drying rate, heating/cooling rate and material sensitivity relationships. The results can be displayed and manipulated in various ways:

ZONAL MODELLING CALCULATIONS

ZONE TYPE = convection drying zone ZONE NUMBER = 3

INPUT PARAMETERS	ABOVE	BELOW	OUTPUT PARAMETERS	
Velocity/ms-1 =	50.00	50.00	Moisture In/%wb =	7.60
Air Temp/C =	210.0	210.0	Moisture Out/%wb =	5.37
Humidity/gg-1 =	0.1800	0.1800	Solid Temp In/C =	70.0
Solvent/%LEL =	0.	0.	Solid Temp Out/C =	97.2
			Fan Energy/kW =	81
Noz to Mat/mm =	10.0	10.0	Fan Volume(m3/hr) =	98098
Noz Pitch/mm =	250.0	250.0	Fuel Energy/kW =	1348
Noz % Area =	1.52	1.52	Prod Quality/Jg-1 =	31.94
Makeup Temp/C =	30.0		Fan Energy(£/yr) =	14584
Make Hum/gg-1 =	0.0150		Fuel Cost(£/yr) =	69020
			Amort Cost(£/yr) =	67674
Zone Length/m =	3.39			
Linespeed/mpm =	900.00		Energy Cost(£/yr) =	83605
ACTION REQUIRED :			Total Cost(£/yr) =	151279

Profiles Next Zone Recalc Zone Inputs/Outputs Quality Help Menu

The above display tabulates the input and output parameters on a zonal basis, together with the operating costs. Any of the inputs can be adjusted directly on screen and the zone remodelled to examine the signif- cance of various sets of inputs upon productivity and operating costs.

MOISTURE AND TEMPERATURE PROFILES FOR ZONE

ZONE NUMBER = 3 ZONE TYPE = convection drying zone

i	Dist /m	Tsolid /C	Top Layer Moist/%wb	Bot Layer Moist/%wb	Overall Moist/%wb	Top Vap Rate /kgm-2h-1	Bot Vap Rate /kgm-2h-1
6	0.	70.0	31.23	3.50	7.60	0.	0.
7	0.13	73.8	30.58	3.75	7.68	47.03	-75.49
8	0.25	76.3	29.61	3.98	7.68	68.63	-70.60
9	0.38	77.8	28.38	4.20	7.63	84.22	-65.81
10	0.50	78.7	26.97	4.40	7.55	93.56	-60.58
11	0.63	79.1	25.41	4.58	7.43	98.69	-55.96
12	0.75	79.3	23.75	4.75	7.30	100.71	-52.01
13	0.88	79.3	22.02	4.91	7.15	100.96	-48.45
14	1.00	79.2	20.22	5.06	7.00	99.97	-45.30
15	1.13	78.9	18.39	5.19	6.85	97.12	-42.69
16	1.25	78.7	16.53	5.32	6.70	94.32	-40.32
17	1.38	78.5	14.67	5.44	6.56	90.26	-38.06
18	1.50	78.6	12.83	5.56	6.42	85.44	-35.64

ACTION REQUIRED :

Profiles Next Zone Recalc Zone Inputs Outputs Help Menu Quit

The profiles of the modelling calculations displayed above present the material moisture, temperature, and evaporation rate from point-to-point along the zone. Only part of the profile is on screen at one time. The whole profile is displayed section by section sequentially by repeat pressing of the Enter key.

```
FACTORIAL OPTIMISATION PROCEDURE WITH AUTOMATIC PARAMETER SCANNING

Eight different parameters can be scanned automatically to assist the
optimisation procedure.  All parameter combinations will be evaluated.
It is recommended that each parameter range be specified carefully to
ensure the total number of combinations is realistic ( number < 500 ).
ZONE NUMBER =     3              ZONE TYPE =  convection drying zone

THE PARAMETER RANGES ARE ENTERED BELOW AS FOLLOWS:
1. to skip parameter, press ENTER (parameter set to default value)
2. parameter range defined by three values - {start},{end},{increment}

Param 1 : Air Velocity (above)/ms-1 :   40.0000  70.0000  10.0000     4
Param 2 : Air Velocity (below)/ms-1 :   40.0000  70.0000  10.0000     1
Param 3 : Air Temperature (above)/C :  100.0000 400.0000  25.0000    13
Param 4 : Air Temperature (below)/C :  100.0000 400.0000  25.0000     1
Param 5 : Air Humidity (above)/gg-1 :    0.0500   0.2500   0.0500     5
Param 6 : Air Humidity (below)/gg-1 :    0.0500   0.2500   0.0500     1
Param 7 : Air Solvent (above)/%LEL  :    0.       0.       0.         1
Param 8 : Air Solvent (below)/%LEL  :    0.       0.       0.         1
NUMBER COMBINATIONS :  260                   ACTION REQUIRED :  r

  Calculations for Zone    Next Zone    Ranges for Parameters    Quit
```

The above display concerns the setting-up of the factorial optimisation search procedure. For each of the eight main airstream parameters a range of values can be specified. The search is simplified if the process involves the same airstream conditions on both sides of the web, single-sided drying, or only water-wetted materials.

USE OF MODELLING SOFTWARE TO UNDERSTAND THE DRYING PROCESS

In order to understand a typical industrial drying operation the following approach is recommended:

STEP 1 - Measure the key processing properties of the substrate and coating materials, such as drying rate curves and equilibrium isotherms, by laboratory tests (ref. 3);

STEP 2 - Establish the performance of the drying system at typical operating conditions via detailed monitoring or by Pilot Plant trials;

STEP 3 - Set up model of drying system by use of fitting procedure which determines constants for the heat transfer correlations appropriate to the nozzles or energy source installed on the dryer

Once the model setting-up procedures have been completed, the predictions of the software provide detailed point-to-point profiles of the main operating parameters along the dryer. These provide valuable insights towards the factors that control and constrain the process.

The use of the modelling procedures can be illustrated by an industrial case-study which involves the drying of a 8 gsm clay coating on 70 gsm paper, 6.5 metres wide, at a linespeed of 900 metres per minute, on an airfloat dryer operated at typical airstream conditions. Profiles predicted by the software are displayed below in Figs. 1 and 2.

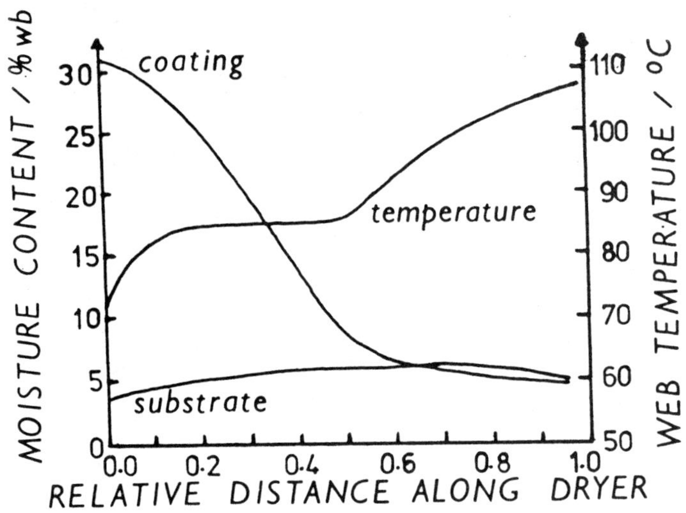

FIG. 1 MATERIAL MOISTURE AND TEMPERATURE PROFILES FOR DRYING OF CLAY COATING ON PAPER.

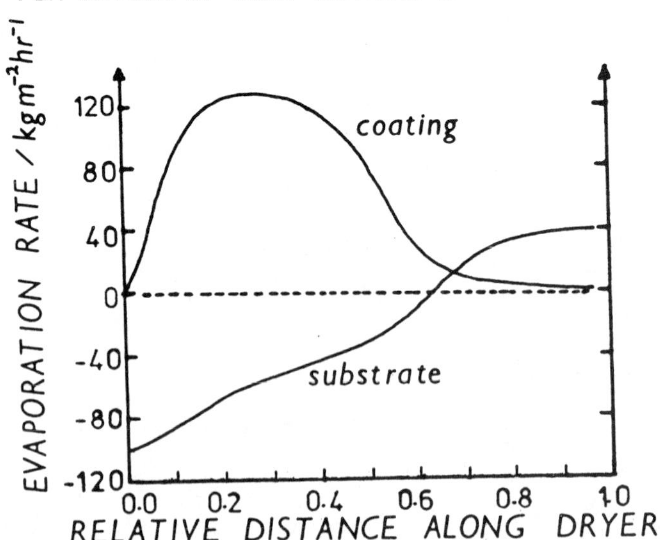

FIG. 2 MOISTURE EVAPORATION PROFILES FOR DRYING OF CLAY COATING ON PAPER.

Note during the first half of the process how the substrate gains moisture which offsets the effectiveness of the top side drying of the coating. The moisture gains may appear small but the problem is amplified because the weight of the substrate is much greater than the dry coating weight as shown by the plot of the evaporation rates. This leads to poor line efficiency during the final 30% of the dryer as it is necessary to redry the substrate to meet the product moisture specification. This demonstrates why there is often considerable scope for enhancing line performance by choosing suitable operating conditions which best match the characteristics of the materials being processed.

USE OF MODELLING SOFTWARE TO EXAMINE PRODUCT QUALITY CONSTRAINTS

The investigation of product quality problems follows a similar approach as discussed in the previous section for understanding the drving process. In addition to the first three steps, a fourth step is required:

STEP 4 - Carry out additional measurements to develop an understanding of the mechanisms which influence the quality sought for the material.

There are many types of quality defects appropriate to drying of coatings, but in this instance we will concentrate upon one of the most common problems of thermal degradation. The same case-study example as used above can illustrate the power of the modelling techniques to investigate such interactions. As thermal degradation is a direct function of the material's heat sensitivity, the thermal history over the drying cycle as shown by the material temperature curve in Fig. 1 will indicate when the quality is most at risk.

As the product temperature is highest during the final stages, a simple criterion for minimisation of thermal degradation of the product is that the material temperature not be allowed to exceed a limiting value. A contour plot is presented on the next page (Fig. 3) which shows the interaction of airstream humidity and airstream humidity upon product temperature at constant nozzle velocity. Note that for each point on this plot the linespeed was adjusted so that the product moisture specification was met at the dryer outlet.

Of particular note is the fact that the shape of the contours vary with the choice of limit product temperature. For temperatures above 100C minima are observed within the contour while at low temperatures maxima mav be seen. By contrast the 95C contour is almost flat.

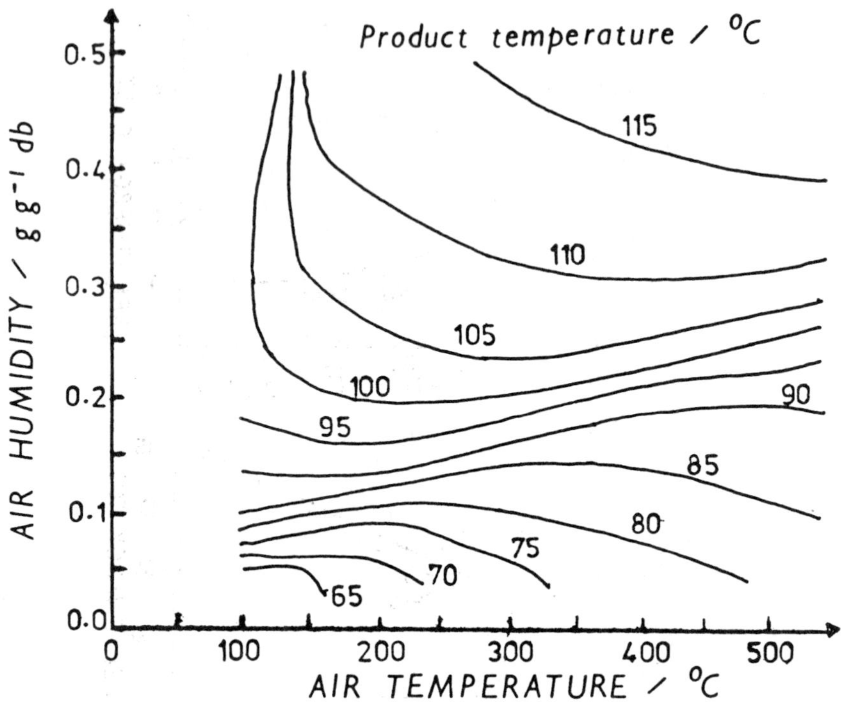

FIG. 3 CONTOURS OF MAXIMUM PRODUCT TEMPERATURE FOR DRYING OF CLAY COATING ON PAPER.

Although the above contours are based upon the simple concept of maximum product temperature, much more complex definitions of product quality can be incorporated into the modelling procedures. For example, the thermal degradation can be modelled via reaction kinetics type expressions such as the Arrhenius rate law where the rate varies with both material temperature and moisture content. Note the use of rate expressions to understand degradation implies that the residence time also becomes an important factor. The author has found this type of approach valuable for optimising linespeeds for applications such as the manufacture of thermal fax papers.

USE OF MODELLING SOFTWARE TO UPGRADE LINESPEED AND PRODUCTION CAPACITY

The use of the modelling software to investigate the production performance of existing plant will involve the first four steps identified earlier. Then via two further steps the scope for increasing linespeed can be determined.

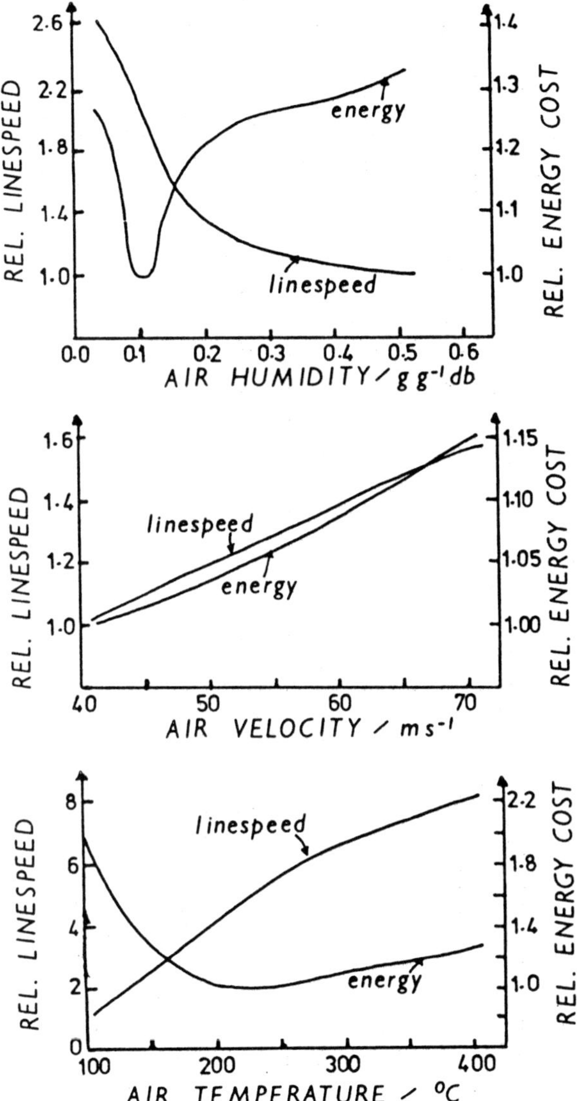

FIG. 4 EXAMPLES OF RELATIONSHIPS BETWEEN PRODUCTION CAPACITY AND DRYER AIRSTREAM PARAMETERS OF VELOCITY, TEMPERATURE AND HUMIDITY.

STEP 5 - Optimise nozzle settings to obtain best heat transfer configuration.

STEP 6 - Use model with factorial parameter scanning procedures to investigate sets of operating conditions which achieve the required product moisture to examine their significance upon line speed.

For modern plant supplied by a reputable manufacturer it is probable that the nozzle settings are close to optimum. However for older plant, it is likely that considerable improvement is feasible.

By contrast, the relationships between capacity and air temperature, humidity and velocity are complex. It is important to realise that the same throughputs can be achieved at quite different combinations of air conditions. This is demonstrated by Fig. 4 which provides examples of the interactions between linespeed, energy cost and airstream conditions. Each of these plots examines the trends with respect to one airstream parameter while the other two are kept constant. It should be noted that significantly different trends can be observed for other pairs of operating variables.

The results obtained by application of the factorial optimisation procedures to the case-study example are summarised on Fig. 5. In this instance it appears there is considerable over-design in the existing installation which suggests in certain circumstances that speed increases of up to 100% may be feasible. It can also be seen that the air humidity/air temperature combinations required are similar regardless of the air velocity chosen.

USE OF MODELLING SOFTWARE TO MINIMISE ENERGY COSTS

For minimisation of the energy costs a similar procedure is followed as in the previous section by use of parametric optimisation studies. The first four steps are the same, but steps 5 and 6 involve a different emphasis :

STEP 5 - Use model with factorial parameter scanning procedures to identify sets of operating conditions which achieve the required product moisture and linespeed;

STEP 6 - Invoke cost functions for fuel for air heating and electricity supply for fans to establish operating regimes with the most favourable total energy costs.

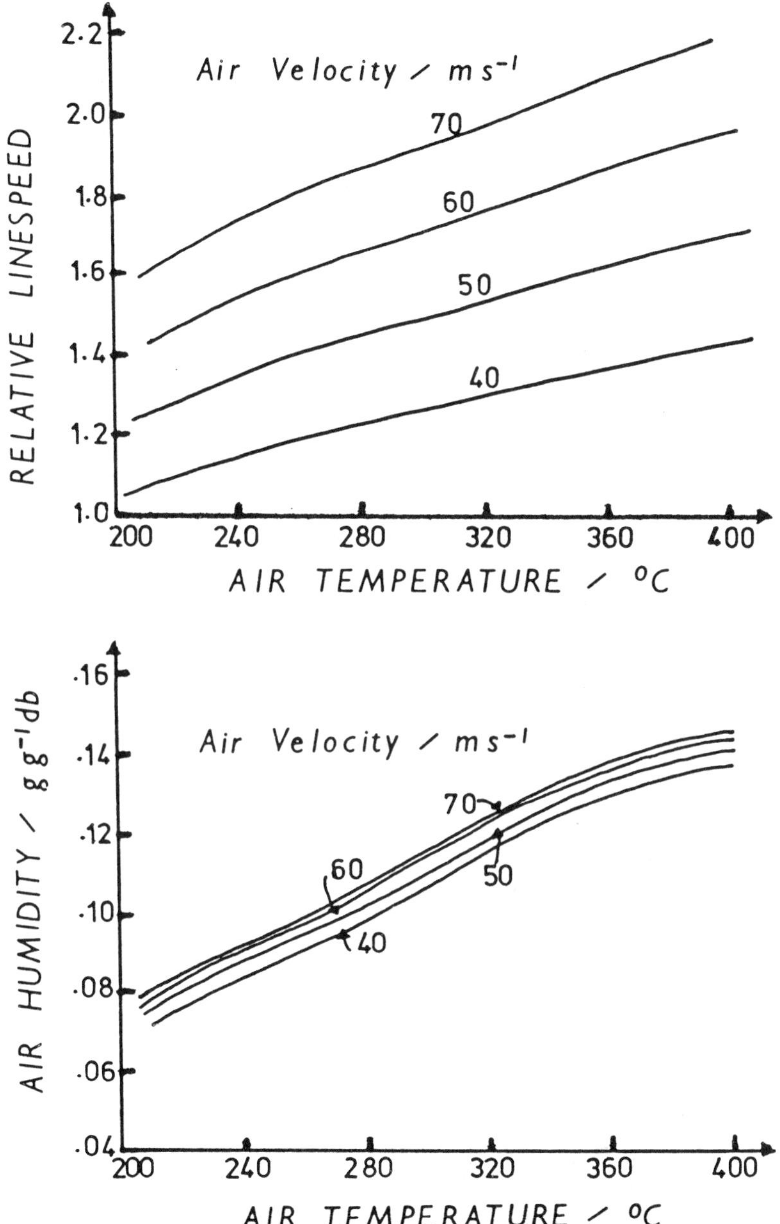

FIG. 5 RESULTS OF FACTORIAL OPTIMISATION ANALYSIS OF SCOPE FOR LINE SPEED ENHANCEMENT FOR CASE STUDY EXAMPLE OF MANUFACTURE OF CLAY COATED PAPER.

The results of an energy cost survey for the case study example are displayed below in Fig. 6. In this instance the airstream was heated by use of direct-fired natural gas at 30 pence per therm. As the present operating conditions for this installation result in an energy cost of £87,000 per year, considerable savings are possible as the best set-up offers costs of less than £60,000 per year. Note these optimised conditions will involve very different and substantially reduced air velocities, humidities and temperatures.

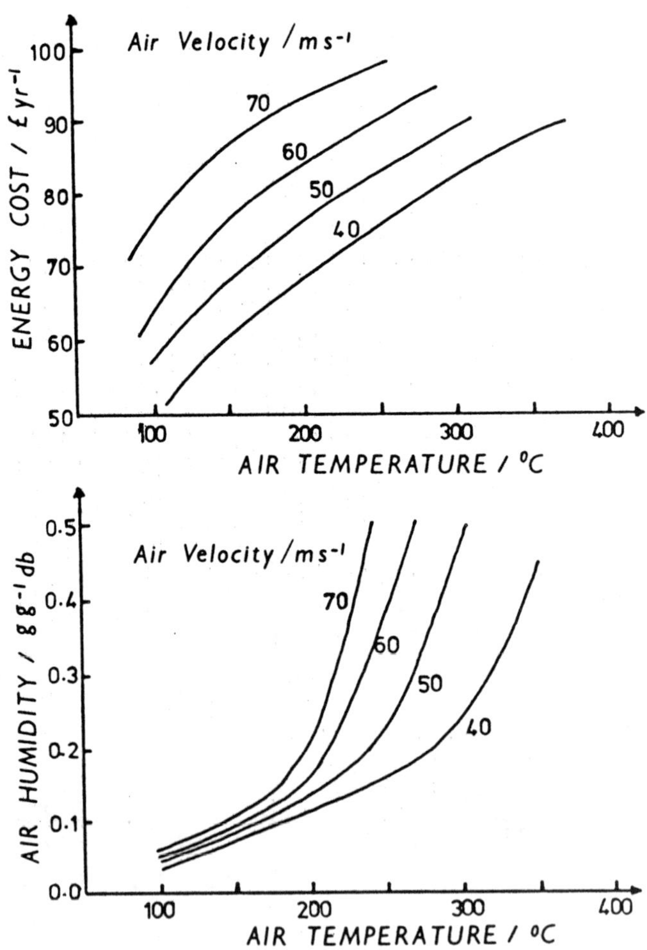

FIG. 6 SIGNIFICANCE OF DRYER OPERATING CONDITIONS UPON ENERGY COSTS FOR MANUFACTURE OF CLAY COATED PAPER

USE OF MODELLING SOFTWARE TO ASSIST DEVELOPMENT OF CONTROL STRATEGIES

A similar procedure is followed as for other situations with the same first four steps and a revised step five:

STEP 5 - Use model with factorial parameter scanning procedures to investigate the significance of upsets in the input conditions upon the main quality parameters.

This type of analysis is particularly valuable to assess the affect of upsets upon the sensitivity of product properties. The upsets may arise from variations of the upstream process such as feed moisture or linespeed, or from problems arising from variation of the airstream conditions.

Fig. 7 plots the contours of product moisture as a function of airstream temperature and humidity assuming the linespeed is kept constant for the example of clay coating of paper. If the product moisture sensor detects a deviation from the setpoint, a chart such as this can be used to help retune the system by identifying the adjustment in airstream conditions needed to correct the product properties. For example, a 0.2% error in moisture can be eliminated typically by a 25 to 30C change in temperature at constant humidity or a 0.050 to 0.075 gg-1 change in humidity at constant temperature.

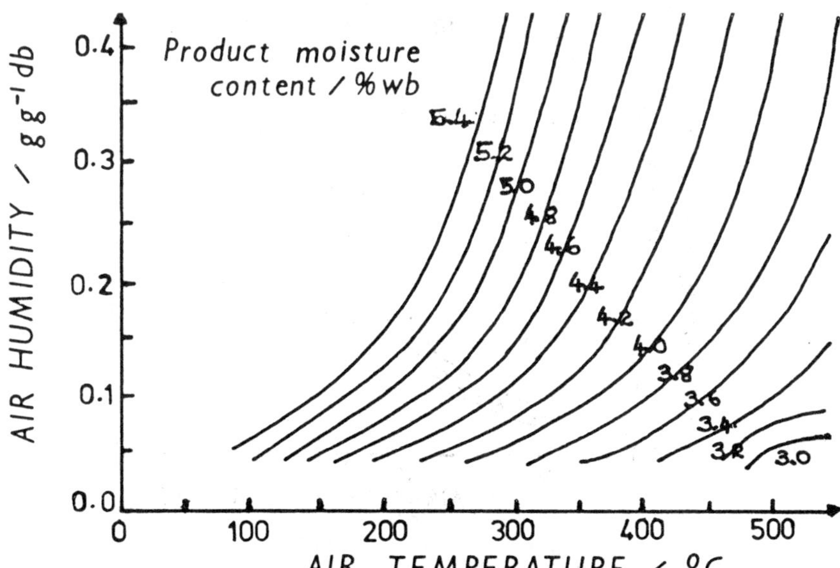

FIG. 7 SENSITIVITY OF PRODUCT MOISTURE CONTROL TO CHANGES IN AIRSTREAM CONDITIONS FOR CASE STUDY EXAMPLE OF CLAY COATING ON PAPER.

Clearly the model offers the scope to greatly enhance the intelligence of the process control system. Computer control systems based upon on-line modelling techniques, such as presented here, are expected in the near future to yield up to an order of magnitude improvement in product quality compared to present technology.

USE OF MODELLING TECHNIQUES FOR THE DESIGN OF NEW PROCESSING LINES

The design of new processing lines involve an extra degree of freedom compared to the examination of performance of existing plant. This is because the size and exact nature of the plant are not fixed in advance. As a result the recommended procedure combines the original four steps with three new steps:

STEP 5 - Use factorial parameter scanning procedures to investigate feasible sets of operating conditions and plant dimensions which give the required product quality;

STEP 6 - Examine capital cost and energy cost functions associated with various plant designs/operating conditions combinations;

STEP 7 - Perform Constraint Analysis to accelerate optimisation procedures and identify most suitable plant design.

The first stage of the design procedure is illustrated by Fig. 8 for the case study example. Although this plot only considers the significance of the air temperature and air humidity parameters, it demonstrates that many alternative designs are possible. These alternatives also reveal that the size of plant and subsequent capital cost can vary substantially.

In practice the design must take proper account of the capital cost, energy costs, product quality and materials of construction constraints as shown in Fig. 9. In this instance the design must lie below the product temperature constraint line obtained from the 100C maximum product temperature contour of Fig. 3. Similarly it can also be shown the design must lie to the left of the materials of construction constraint, to the right of the capital cost constraint, and within the region enclosed by the energy costs constraint contour. This means the actual optimum design is bounded by the much smaller region ABCDE.

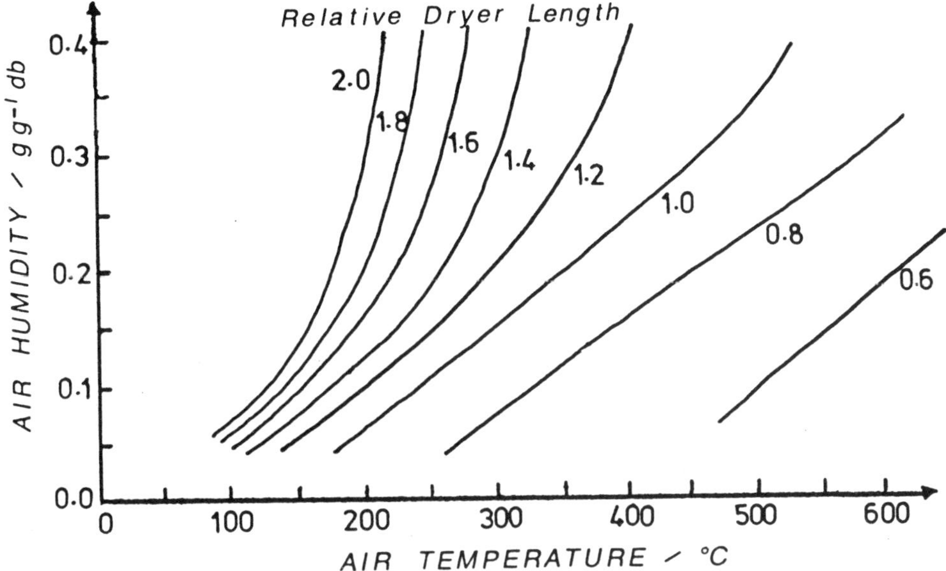

FIG. 8 SIGNIFICANCE OF AIRSTREAM CONDITIONS UPON DRYER SIZE FOR CASE STUDY EXAMPLE OF MANUFACTURE OF CLAY COATED PAPER.

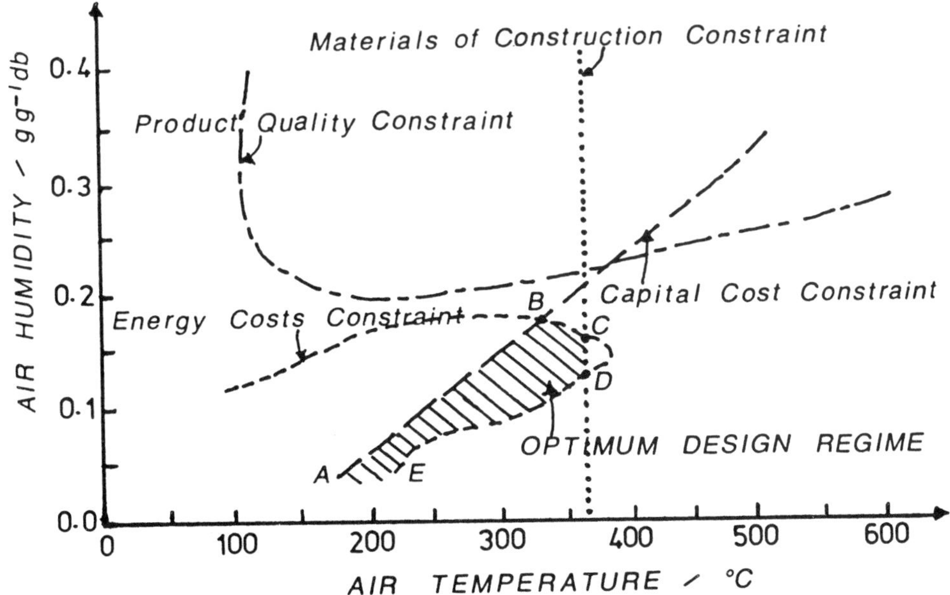

FIG. 9 USE OF CONSTRAINT ANALYIS TO IDENTIFY OPTIMUM DESIGN REGIME FOR DRYER TO MANUFACTURE CLAY COATED PAPER.

The results of the design procedure illustrate the power of the modelling procedures to evaluate a much wider range of designs than is feasible by conventional design techniques. Of equal importance is the supporting role provided by the constraint analysis which ensures the plant design is well matched to the material characteristics and the financial objectives of the project. It is the authors experience that such an approach typically leads to capital cost savings of the order of 10% and energy cost savings of 15% over the lifetime of the plant.

CONCLUSIONS

The examples of the use of the modelling software presented above have demonstrated the power of the techniques to provide detailed insights into factors associated with plant operation. All the graphs presented are based upon an industrial case-study associated with manufacture of clay coated paper. However it should be noted that the trends observed will be unique to this particular installation. For other applications quite different results will be obtained because of the complex interactions arising from the material processing characteristics, energy costs, site factors, nature of plant and the quality factors sought for the product.

REFERENCES

1. J.C. Ashworth - 'The Scientific Approach to Solids Drying Problems', Vols. 1 and 2, Drying Research Ltd., 1982.

2. J.C. Ashworth - 'Using the computer to optimise the dryer', Paper Technology, January 1991, pp. 24-27.

3. J.C. Ashworth - 'Understand your drying process and enhance productivity', The Chemical Engineer, July 1989, pp. 20-24.

Numerical Simulation of Laminar Film Drying

F. Durst, D. Stadler, and H.-G. Wagner
LEHRSTUHL FÜR STRÖMUNGSMECHANIK DER UNIVERSITÄT ERLANGEN-NÜRNBERG, D-8520 ERLANGEN, GERMANY

1 INTRODUCTION

Dryers are important components of any coating installation and the performance of the drying equipment has a strong impact on product quality. For economical reasons, highly efficient drying processes are favoured. However, fast and aggressive drying can have negative effects on the resulting coatings. Hence, the choice of the dryer's operation conditions is limited by the occurrence of drying defects in the coating. Controlling fluid dynamic processes in the gas phase, near the interface to the wet film as well as the flow and diffusion processes within the wet film itself, are crucial for the final film-quality. Numerical simulations of the fluid flow and of heat and mass transfer are useful to determine the critical drying path, which gives highly efficient drying rates together with an optimum coating quality.

Global balances provide information on the total energy requirement of the drying process. An improved, but still coarse picture of the process can be gained, if pre-assumed local heat and mass transfer coefficients can be used to compute the predictions. In reality, there is a strong interrelationship between the processes in the gas phase and those within the wet film. This can be considered by solving the governing equations in the gas and the liquid phase simultaneously. In practical applications the dimensions of the dryer's gas domain are much larger than the vertical extensions of the wet film. This has implications on the numerical treatment of the drying process. Even as computer power increased tremendously during the last decade, storage requirements to sufficiently resolve the areas of interest are extremely high and usually prevent numerical simulation from being used on extensive parameter studies. To circumvent this obstacle, many researchers tend to look only at the phase which they think to be most important for their studies, and then include the influence of the second phase in the boundary conditions. The understandings of the physical background will be improved in this way and the influence of various parameters can be studied. However, this is not sufficient for detailed design of a complete drying process.

In this paper a method is presented which will simulate the processes in both phases simultaneously and provides the necessary resolution at the critical areas named above.

2 GOVERNING EQUATIONS

Conservation Equations

Air flow and the wet film were simulated to predict the dryer performance. The flows considered are assumed to be laminar, steady, and incompressible, Newtonian fluids. They are governed by the following conservation laws for global mass, momentum, energy, and mass of components involved:

mass:

$$\frac{\partial}{\partial x_i}(\rho U_i) = 0 \tag{1}$$

momentum:

$$\frac{\partial}{\partial x_j}(\rho U_i U_j) = -\frac{\partial p}{\partial x_i} + \frac{\partial}{\partial x_j}\tau_{ij} + S_U \tag{2}$$

where

$$\tau_{ij} = \mu\left(\frac{\partial U_i}{\partial x_j} + \frac{\partial U_j}{\partial x_i}\right) - \frac{2}{3}\mu\frac{\partial U_l}{\partial x_l} \tag{3}$$

Here ρ is the fluid density, U_i are the cartesian velocity components, x_i are the cartesian coordinates, p is the pressure, τ_{ij} means the stress tensor and μ stands for the dynamic viscosity. Additional forces, such as lift forces caused by density gradients are included in the source term S_U on the right hand side.

enthalphy :

$$\frac{\partial}{\partial x_j}(\rho U_i c_v T) = -\frac{\partial}{\partial x_i}\dot{q}_i - T\left(\frac{\partial p}{\partial T}\right)_v \frac{\partial U_i}{\partial x_i} - \tau_{ij}\frac{\partial U_i}{\partial x_j} + S_h \tag{4}$$

where

$$\dot{q}_i = \lambda\left(\frac{\partial T}{\partial x_i}\right) \tag{5}$$

Here T means temperature, c_v is the heat capacity, q_i stands for the heat flux and λ denotes heat conductivity. The source term S_h includes heat sinks caused by evaporation.

concentration:

$$\frac{\partial}{\partial x_j}(\rho U_i C_k) = -\frac{\partial}{\partial x_i}\dot{m}_i + S_{C_k} \tag{6}$$

where

$$\dot{m}_i = \rho D_k\left(\frac{\partial C_k}{\partial x_i}\right) \tag{7}$$

Here C_k is the concentration, m_i means mass flux and D_k indicates the diffusivity. For each evaporating component of the wet film, a concentration equation has to be solved.

In case of two volatile components evaporating into the gas domain, a system of six coupled partial differential equations results for the gas domaine and for the wet film. These equations will be solved numerically. They are coupled by the boundary conditions at the free surface of the wet film.

Boundary conditions

The following boundary conditions apply:

at the wall, no-slip condition for velocities, adiabatic wall, and no loss of matter are defined:

$$u_t = 0 \, , \qquad u_n = 0 \, , \qquad \frac{\partial t}{\partial n} = 0 \, , \qquad \frac{\partial c}{\partial n} = 0 \; ; \tag{8}$$

at the inlet, velocities, temperature, and concentrations are specified:

$$u_i = u_{i\ inlet} \, , \qquad t = t_{inlet} \, , \qquad c_i = c_{i\ inlet} \; ; \tag{9}$$

at the exit, zero gradient in flow direction for all dependent variables is assumed:

$$\frac{\partial u_i}{\partial n} = 0 \, , \qquad \frac{\partial t}{\partial n} = 0 \, , \qquad \frac{\partial c_i}{\partial n} = 0 \; ; \tag{10}$$

at the web surface, no-slip condition, no loss of matter are defined:

$$u_t = 0 \, , \qquad u_n = 0 \, , \qquad \frac{\partial c}{\partial n} = 0 \; ; \tag{11}$$

the temperature or the heat flux can be specified:

$$t_{film} = t_{web} \, , \qquad \dot{q}_{film} = \dot{q}_{web} \; ; \tag{12}$$

at the interface, heat and mass flux, shear stress and velocities are defined:

$$u_{i\ film} = u_{i\ gas} \, , \quad \tau_{nt\ film} = \tau_{nt\ gas} \, , \quad t_{film} = t_{gas} \, , \quad \dot{q}_{film} = \dot{q}_{gas} \, , \quad \dot{m}_{film} = \dot{m}_g \tag{13}$$

The concentration of the solvents at the wet film surface was calculated according to

$$C_{1\ Surface} = \frac{1}{1 + \frac{M_2}{M_1}\left(\frac{p_\infty}{p_S} - 1\right)} \tag{14}$$

$C_{1\,Surface}$ Concentration of Solvent 1 at the wet film surface
M_1, M_2 Molecular weight of Solvent 1 and Gas
p_∞ Pressure within the gas domain
p_S Saturation pressure of solvent 1 for the surface temperature

Calculation of properties

Equations (1) to (7) require property data. Values for viscosity, density, heat capacity, and heat conductivity have to be specified. These values can be determined as a function of temperature for all species with the mean value obtained according to the weight fraction [1]. The gas phase is treated as an ideal gas. The diffusivity values of the solvents in the gas phase are calculated according to [2]:

$$D_{12} = \frac{1.013 \cdot 10^{-7}\ T^{1.75} \left(\frac{M_1 + M_2}{M_1 M_2}\right)^{0.5}}{p\left[\left(\sum v_1\right)^{\frac{1}{3}} + \left(\sum v_2\right)^{\frac{1}{3}}\right]^2} \quad \textbf{(15)}$$

D_{12} - binary diffusion coefficient [m^2/s]
T - temperature [K]
p - pressure [bar]
M_i - molecular weight [g/mol]
v_i - specific volume

The values for the diffusivity of the solvents in the polymer-solvent mixture within the wet film are calculated according to Duda [3]:

$$D_1 = D_{01} \exp\left[-\frac{w_1 \hat{V}_1^* + w_2 \zeta \hat{V}_2^*}{V_{FH/\gamma}}\right](1 - 2\chi\phi_1)(1 - \phi_1)^2$$

$$V_{FH/\gamma} = w_1\left[\frac{K_{11}}{\gamma}\right](K_{21} + T - T_{g1}) + w_2\left[\frac{K_{12}}{\gamma}\right](K_{22} + T - T_{g2}) \quad \textbf{(16)}$$

$\hat{V}_1^*$ - specific critical hole free volume of pure solvent
$\hat{V}_2^*$ - specific critical hole free volume of polymer
T - temperature of mixture
T_{gi} - glass transition temperature of component i
K_{ij} - free volume parameters
ζ - ratio of critical molar volume of solvent to critical molar volume of polymer
γ - overlap factor for free volume
w_1 - mass fraction of solvent
w_2 - mass fraction of polymer
ϕ_1 - volume fraction of solvent
D_{01} - preexponential factor
χ - interaction parameter of Flory-Higgins theory

3 THE NUMERICAL METHOD

Discretization and Solution Algorithm

The integral form of Eqs. (1) to (7), obtained by integrating over a control volume and employing Gauss theorem, allows for a straightforward finite volume discretization. Detailed information on the finite volume method is provided by Perić [4].

The solution domain is decomposed into subdomains, or blocks which are solved independently, and are coupled by their boundary values. The solution domain is bounded by dryer walls, its gas inlet, its gas outlet and the surface of the web material. In our study we make only use of two blocks, one for the air space and one for the wet film. These blocks are subdivided into a finite number of contiguous quadrilateral control volumes as shown in Figure 1. At a block's border, one control volume may face more than one neighbouring control volume as shown in Figure 2. Thus, a local refinement of the grid can be accomplished. The integrated equations are linearized and values of dependent variables at locations of the control volume face are expressed through nodal values by means of linear interpolation (central differences).

For the solution domain as a whole, a system of algebraic equations with a five-diagonal coefficient matrix results, which is solved by the strongly implicit procedure of Stone [5]. The solution algorithm is based on the SIMPLE method of Patankar and Spalding [6]. The equations are alternately solved on the subdomains in an iterative manner, so that the boundary values at the interface are taken from the neighbouring block. For each subdomain the conservation equations are solved sequentially. Temperature and concentration dependent properties, like viscosity, density, diffusivity, heat capacity, heat conductivity, and saturation pressure are updated. The whole process is repeated until convergence is obtained. For more details, see the thesis of Stadler [7], and Perić et al. [8].

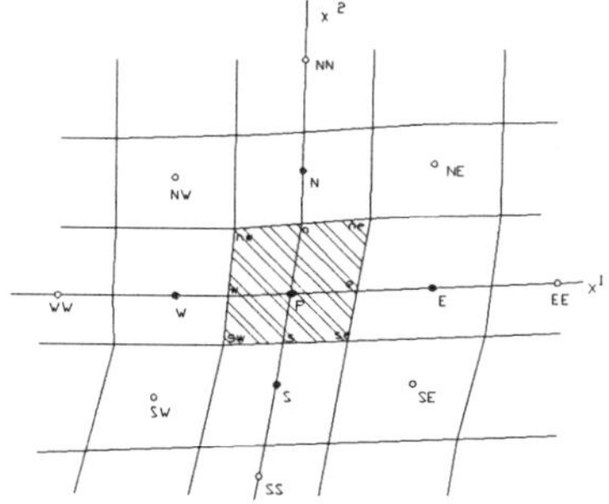

Figure 1: Control volume with neighbouring volumes.

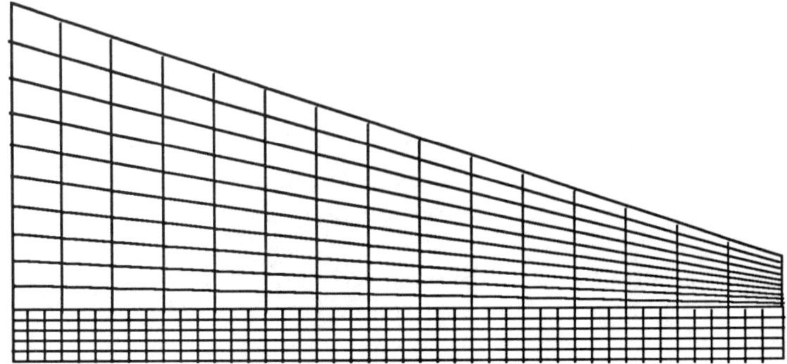

Figure 2: Numerical grid of a laminar dryer before calculation.

4 THE LAMINAR DRYER AND TEST CASES

Basic geometry

Figure 3 shows the basic geometry of the laminar dryer as considered in this study. Its design was suggested by Durst [9] to prevent coating defects during the initial drying process. The working principle of this dryer is also explained by [10] and [11].

Hot air is fed in through the porous top wall and exits on the right side. The inflow can vary along the top wall, so that the condition for laminar flow is always satisfied in the vicinity of the wet film. The coated web enters at the lower end of the left wall and moves at the bottom in parallel to the lower wall. The walls are assumed to be adiabatic. Heat transfer and evaporation of solvents occurs at the interface between wet film and air. The wet film thickness decreases according to the loss of solvent due to evaporation. Heat transfer from the air to the backside of the web is taken into account, while additional heating from the backside of the web can be included. The length of the dryer used is *3.00 m*, exit height is *0.1 m*, and the inclination of the top wall is *9.4°*.

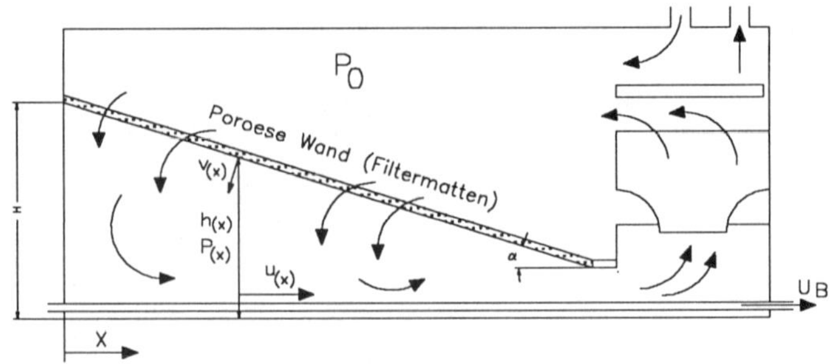

Figure 3: Basic geometry of the laminar dryer.

Test cases

The wet film consisted of a two-solvent system made of tetra-hydro-furan (THF) and methyl-ethyl-ketone(MEK)and the polymer poly-vinyl-acetate (PVA).

Gas (Air)	**case 1**	**case 2**	**case 3**	**case 4**
flow rate	1.5 m^2/s	1.5 m^2/s	1.5 m^2/s	1.5 m^2/s
temperature	40 °C	60 °C	80 °C	80 °C
conc. MEK	0	0	0	0
conc. THF	0	0	0	0
add. heating	2000 W/m^2	2000 W/m^2	2000 W/m^2	500 W/m^2

Wet film	**case 5**	**case 6**	**case 7**	**case 8**	**case 9**	**case 10**	**case 11**
web velocity	1 m/s	1 m/s	1 m/s	1 m/s	1 m/s	1 m/s	1 m/s
film height	100 μm	100 μm	100 μm	100 μm	100 μm	100 μm	100 μm
temperature	20 °C	20 °C	20 °C	20 °C	20 °C	20 °C	20 °C
conc. MEK	20 %	30 %	40 %	50 %	60 %	0 %	60 %
conc. THF	60 %	50 %	40 %	30 %	20 %	60 %	0 %

Table 1: Input values of test cases.

5 RESULTS

Numerical Grid

Calculations for dryer geometry as shown in Figure 3 were performed on a grid that consisted of *300* by *5* control volumes for the wet film domain. The grid for the gas domain was refined in four steps from 30 by 25 to 300 by 25 control volumes. Figure 2 shows a numerical grid at the start of the calculation. During the calculation the program self-adjusts the grid height according to the decrease of the wet film thickness, that is due to evaporation of the solvents. The figure shows the vertical dimensions of the two domains scaled separately to improve the visibility.

Velocity profiles

Figure 4 shows the predicted velocity vectors. Velocity profiles show a steep decrease close to the surface of the coating where air velocity is identical to the wet films velocity. The influence of the shear force at the free surface on wet film velocity is negligible, while the velocity within the wet film is nearly identical to the web speed. Above the wet film a very thin boundary layer is formed. Its vertical extension decreases as long as the flow is accelerated as opposed to the constant flow over a flat plate, where the boundary layer increases furthermore and becomes turbulent.

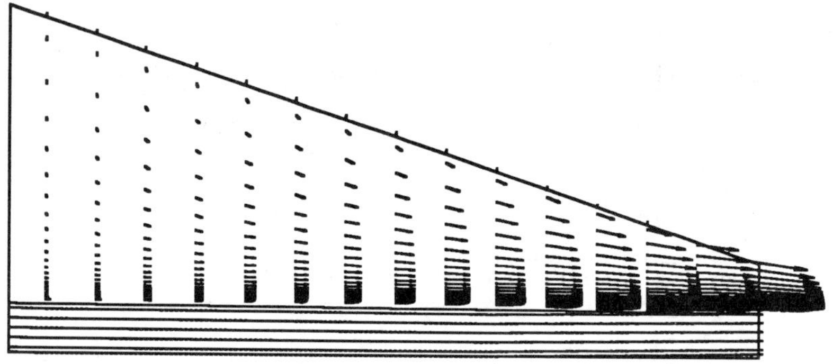

Figure 4: Predicted velocity vectors.

Streamlines

The streamlines in Figure 5 indicate the direction of the main flow. The air enters through a porous wall on the top. Air flow turns and runs in parallel close to the free surface of the wet film.

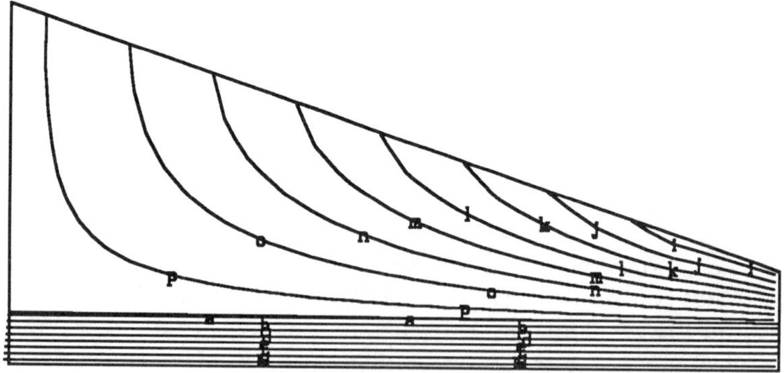

Figure 5: Predicted streamlines.

Temperature distribution

Figure 6 shows the temperature distribution within the wet film and the air boundary layer for the combination of test cases 4 and 9. Vertical scaling of the air domain is further enlarged for better visibility. The shape of the temperature boundary layer is similar to the air boundary layer. Heat transfer improves as the boundary layer thickness decreases, and the drying gradient gets steeper. Temperature has its minimum in vertical direction at the free surface where the solvents' evaporation takes place. Figure 7 shows the temperature at the surface of the wet film. Temperature drops rapidly at the entrance where heat and mass transfer are at their maximum. As soon as the boundary layer gets saturated with the solvents temperature rises again, due to the reduced evaporation rate. A further decrease in the boundary layer thickness again improves the evaporation and temperature decreases.

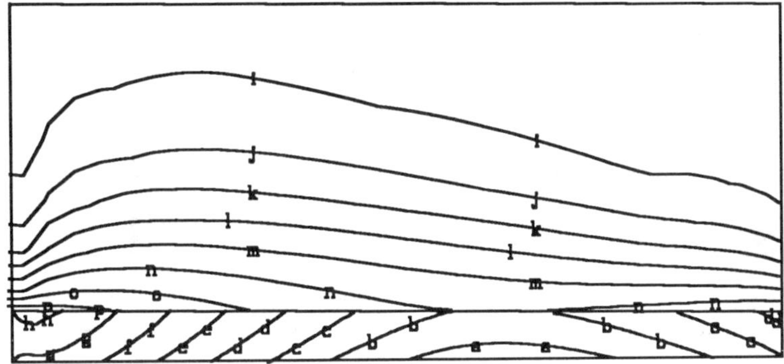

Figure 6: Temperature distribution within film and air boundary layer.

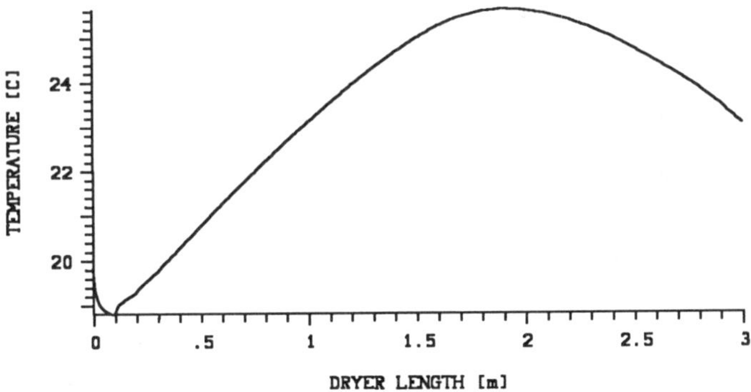

Figure 7: Surface temperature of the film.

Concentration distribution

Figure 8 shows the concentration distribution of the solvent MEK within the wet film and the air boundary layer. The lowest values of solvent concentration in vertical direction within the wet film is at the free surface. This figure shows that the solvent content of the wet film is the smaller, the closer you get to the free surface. Hence, the solvents' diffusivity is reduced in such regions, and the drying process is slowed down.

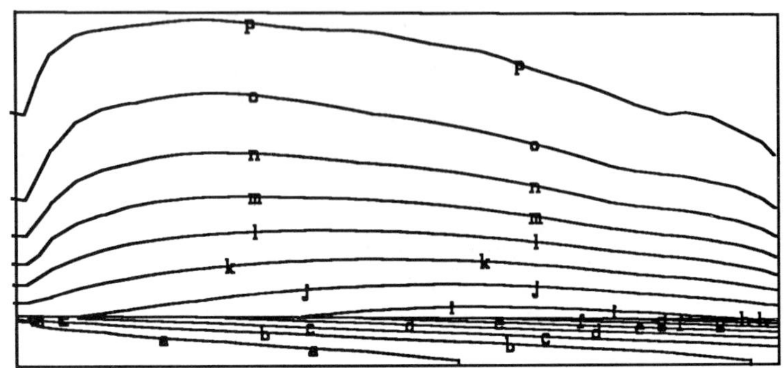

Figure 8: Concentration distribution of MEK within film and boundary layer.

Wet film thickness

Figure 9 shows the progress of drying indicated by reduction of the wet film thickness. In the case we considered the solvent content is reduced by 15 %.

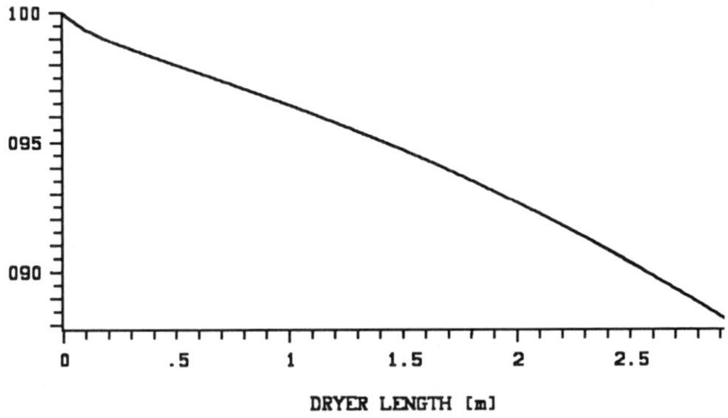

Figure 9: Decrease of the film height due to evaporation.

Monitoring critical values

Figure 10 shows the shear stress between the air flow and the film surface. The value of the shear stress increases with rising air velocity. Figure 11 denotes the distribution of the viscosity at the film surface. In order to prevent from wave-formation on the film surface, viscosity must be increased for higher shear stress.

It is not necessary to calculate the appearance of wave-formation on the film surface, but it is sufficient to know the criterions for wave-formation or instabilities. Eg. one of these criterions can be the viscosity.

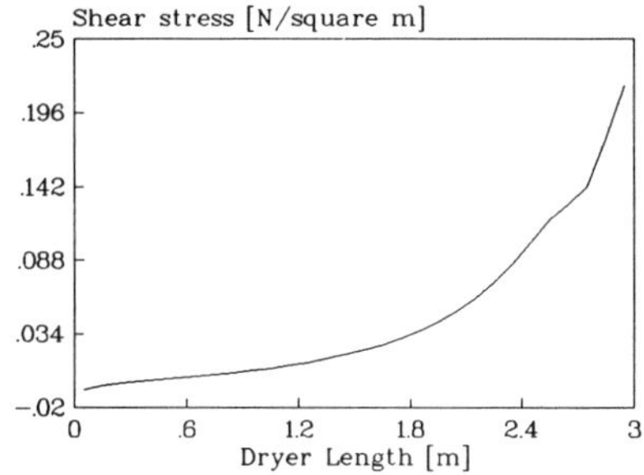

Figure 10: Shear stress acting on the film surface.

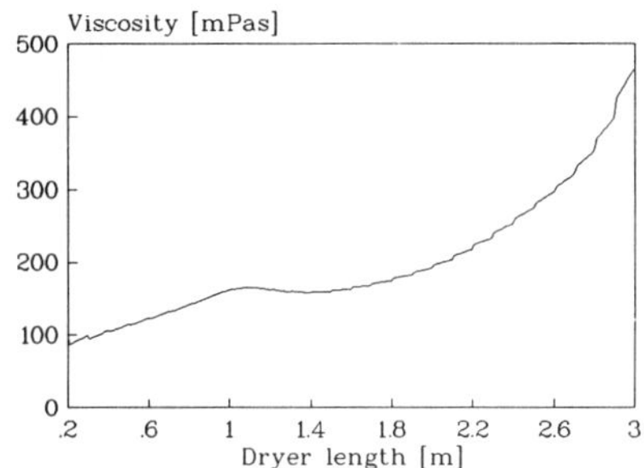

Figure 11: Viscosity of the film at the surface.

Comparison of results from different grids

The figures given in Table 2 demonstrate the influence of the number of control volumes on computation time. They strongly suggest, to do the calculations using less control volumes in the air domain. For those coarser grids the solutions are nearly as exact as they would be using finer grids.

Calculations on coarser grids within the film domain could not be performed, due to the bad aspect ratio. In general it is very difficult to get solutions on grids with a aspect ratio greater than 500.

Grid size of gas domain	300 x 25	60 x 25	30 x 25	20 x 25
Film temperature at exit	288.78 K	290.2 K	290.75 K	290.88 K
Film height at exit	92 %	91 %	90.5 %	90.5 %
Computational time	17,261 s	1,317 s	498 s	330 s

Table 2: Influence of grid size on computational time.

6 CONCLUSIONS AND FURTHER REMARKS

- Numerical simulations provide improved understandings of thepicture of processes in the boundary layer and within the wet film.

- The dimensions of a dryer unit are usually much larger than the thickness of the wet film. The proposed numerical procedure delivers sufficient spatial resolution at the critical areas by the use of local refinement.

- Improved performance is achieved through reduction of the control volumes's aspect ratio in the wet film despite the increase in number of control volumes.

- Heat and mass transfer rates are calculated at the interface to predict the progress of drying. The resulting information is necessary to design and control the drying process by changing the gas flow, or through the inclusion of additional heat sources at the backside of the web.

- Monitoring critical values is possible to predict onset of instabilities, but spacial resolution is not yet sufficient to simulate convection cells.

- Our program is useful to verify programs based on simpler drying models and to calculate local heat and mass transfer coefficients, that again can serve as an input for those programs.

- The major shortcomings of numerically simulating drying processes today is the lack of reliable data on the component's physical properties and their interaction within the mixture. Some of the existing models with the aim of predicting the behaviour of the solvent systems during drying show good agreement with experimental data. Nevertheless, a combined approach of theoretical, experimental and numerical methods is necessary to optimize drying processes.
The method shown can also be applied to turbulent flows. A turbulence model has to be used to describe the influence of turbulence on the flow and on transporting processes, but storage requirements and computing time increase even more.

REFERENCES

[1] Reid, Prausnitz, Sherwood, 'Properties of gases and liquids´, 3rd. edition, 1977.

[2] VDI-Gesellschaft für VT u. CIW (GVC), 'VDI-Wärmeatlas´, VDI Verlag GmbH, Düsseldorf, 5th edition, 1988.

[3] Duda J.L., Transport Properties in 'Devolatilization of Polymers´, Section III, pp. 89-129, Hanser Publishers, Munich, 1983.

[4] Peric M., 'A Finite Volume Method for the Prediction of Three-Dimensional Fluid Flow in Complex Ducts´, Ph.D Thesis, University of London, 1985.

[5] Stone H.L., 'Iterative Solution of Implicit Approximations of Multi-Dimensional Partial Differential Equations´, SIAM J. Num. Anal., vol. 5, pp. 530-558, 1968.

[6] Patankar S.V., Spalding D.B., 'A Calculation Procedure for Heat, Mass and Momentum Transfer in Three-Dimensional Flows´, Int. J. Heat Mass Transfer, vol. 15, pp. 1787-1806, 1972.

[7] Stadler D., 'Numerische Simulation der Trocknung von beschichteten, bahnförmigen Materialien´, Diploma Thesis, Lehrstuhl für Strömungsmechanik, University Erlangen-Nürnberg, 1991.

[8] Perić M., Schäfer M., and Schreck E., 'Computation of fluid flow with a parallel multigrid solver´, in Proceedings of the conference on parallel computational fluid dynamics, (Stuttgart, Elsevier, Amsterdam), 1991.

[9] Durst F. et al.: "Process and Device for Drying a Liquid Layer Applied to a Moving Carrier Material", US Patent 4,999,927, 1991.
[10] Durst F., Lange U., Raszillier H., Wagner H.-G., 'Performance study of laminar flow dryer for applications in film coating´, to appear in Proceedings of the IS&T's 44th Annual Conference, St. Paul, MN, 1991.
[11] Durst F., Stadler D., Wagner H.-G., 'Numerical study of a laminar flow dryer´, presented at 1992 AIChE National Spring Conference, New Orleans, LA, 1992.

Air Flotation Drying and Non-contact Web Handling

R.T. Proctor
SPOONER INDUSTRIES LIMITED, RAILWAY ROAD, ILKLEY,
WEST YORKSHIRE LS29 8JB, UK

1 INTRODUCTION

As the market place demands higher and higher specification products so the onus is placed on today's paper maker and converter to handle the ever-increasing variety and sophistication of coatings and impregnations necessary for the production of such products.

The demand for these higher specification products has forced the development of advanced heat transfer and contactless web handling systems.

The object of this Paper is to highlight the "State of the Art" of Flotation and Web Handling apparatus and illustrate the principles upon which these inventions were developed.

To this end the inventions to be discussed will be:

1. Air Flotation Dryers
2. Air Turn Roll
3. Jet Foil$_{TM}$ Cylinder

In essence each of the inventions was taken from conception to manufacture with a discreet set of objectives in mind. However, it will become apparent that they can be brought together in a variety of possible combinations to achieve today's hi-tech finished products.

The pressure behind these developments stemmed originally from the industry's desire to run at higher operating speeds combined with higher specification coatings and improved product quality.

2 AIR FLOTATION DRYERS

Prior to the introduction of modern air float drying techniques, a conventional dryer involved the impingement of air at high velocity on to the top surface of the coated web, the web being supported directly by rollers (Fig. 1) or, alternatively, carried by a permeable fabric conveyor (Fig. 2). The drawbacks to this system are that the web, or carrying conveyor, tends to sag between the rollers and unless hot air is introduced to the base, sweating may occur on the underside. Without a conveyor the air velocity is limited, as high velocity induces flutter unless a high web tension is used.

Conversely if the dryer does have a fabric conveyor it has the tendency to mask the heat transfer effects on the underside of the sheet, thus reducing the drying capability. Of course frequent roller maintenance and contamination, or marking of the product by the conveyor, can also be a problem.

Thus the route was open for the development of a system that would dry the product evenly and transport it in a contactless mode.

In its simplest form air float drying has been servicing the industry for many years.

Spooner Industries have been manufacturing forced convection drying equipment using nozzle jet impingement techniques since 1932 and the founder, William Wycliffe Spooner, held all the original patents on this form of drying.

Air flotation dryers were first introduced by Spooner in around 1961 and since that date Spooners have adopted a policy of continual ongoing product development to capitalise on the basic foundation techniques.

The flotation dryer of today, manufactured by Spooner Industries, utilises advanced aerodynamic effects brought about by the design of the flotation air bar, the geometry of which has been arrived at through an in-depth knowledge and appreciation of the air movement in this area, coupled with an understanding of the process requirements.

However, that is only half the story. Because quality is paramount at Spooner, the rigorous standards laid down as foundations by this development work demands a high level of skill in the manufacture of these products as well as a high investment in specialist tooling and equipment.

Pressure Chamber
Coating
Paper Web
Support Rollers
RE-ENTRANT NOZZLE SYSTEM
WITH WEB CARRIED ON ROLLS

Fig 1

Pressure Chamber
Coating
Paper Web
Glide Boxes
Conveyor Support Rollers
RE-ENTRANT NOZZLE SYSTEM
WITH WEB CARRIED ON FABRIC CONVEYOR

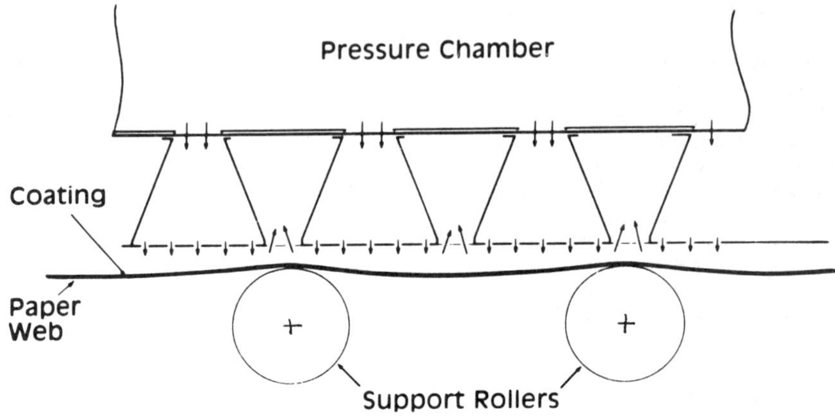

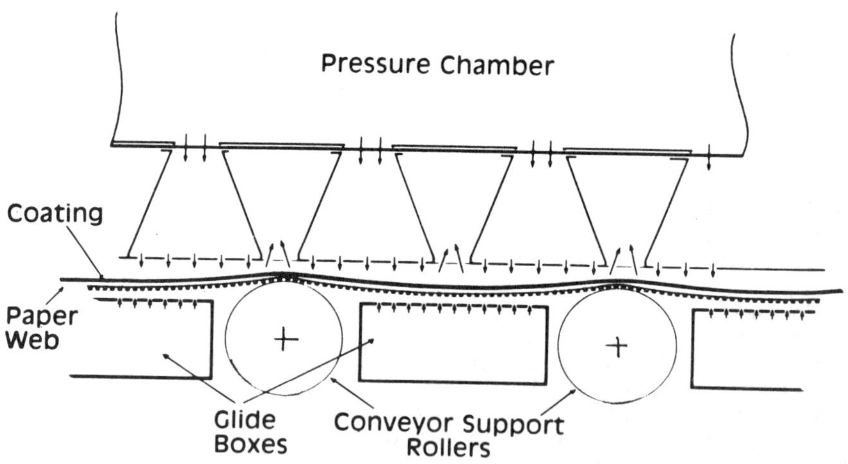

Fig 2

3 THE SPOONER AIR BAR

The foundation stone of Spooner Air Flotation Technology is the air bar.

Over the years Spooner have developed a family of air bars to cater for the varying demands of today's market.

Spooner Float (Fig. 3)
Film, Fax Papers, Release Papers, Clay Coated Papers etc.

SupaFloat (Fig. 4)
Aluminium Strip, Non-Wovens etc.

Max Float (Fig. 5)
Steel Strip, (Dryer currently operating with 10 kg/m^2)
Non-Wovens etc.

These categories are by no means fixed and each case and subsequent choice of air bar is assessed on its own merit with reference to Spooner's experience and test work conducted in the Hi-Tech Spooner Test Centre, thereby ensuring optimum air bar selection.

The "work horse" and flagship of the Spooner air bar fleet without doubt is the Spooner Float and it is this air bar that this paper will concentrate on.

Successful flotation and drying of continuous webs can, at first glance, appear to present somewhat of a paradox in that it requires different performance characteristics i.e.

1. A very stable cushion required to maintain web stability.

2. A high degree of air turbulence at the web surface to generate high rates of heat transfer.

It is worth noting that nozzle systems that don't conform to these two characteristics can exhibit very serious flaws in that they may give stable web support but the heat transfer rates achievable are low and hence drying is slow. Conversely achievable drying rates may be high but flotation is sensitive to, and dependent upon, external conditions such as web tension, web weight and nozzle velocity.

The key to the success of the air bar lies in the geometry of the upper surface. Through a long and continuous development process each discreet dimension and the complete assembly was optimised for performance,

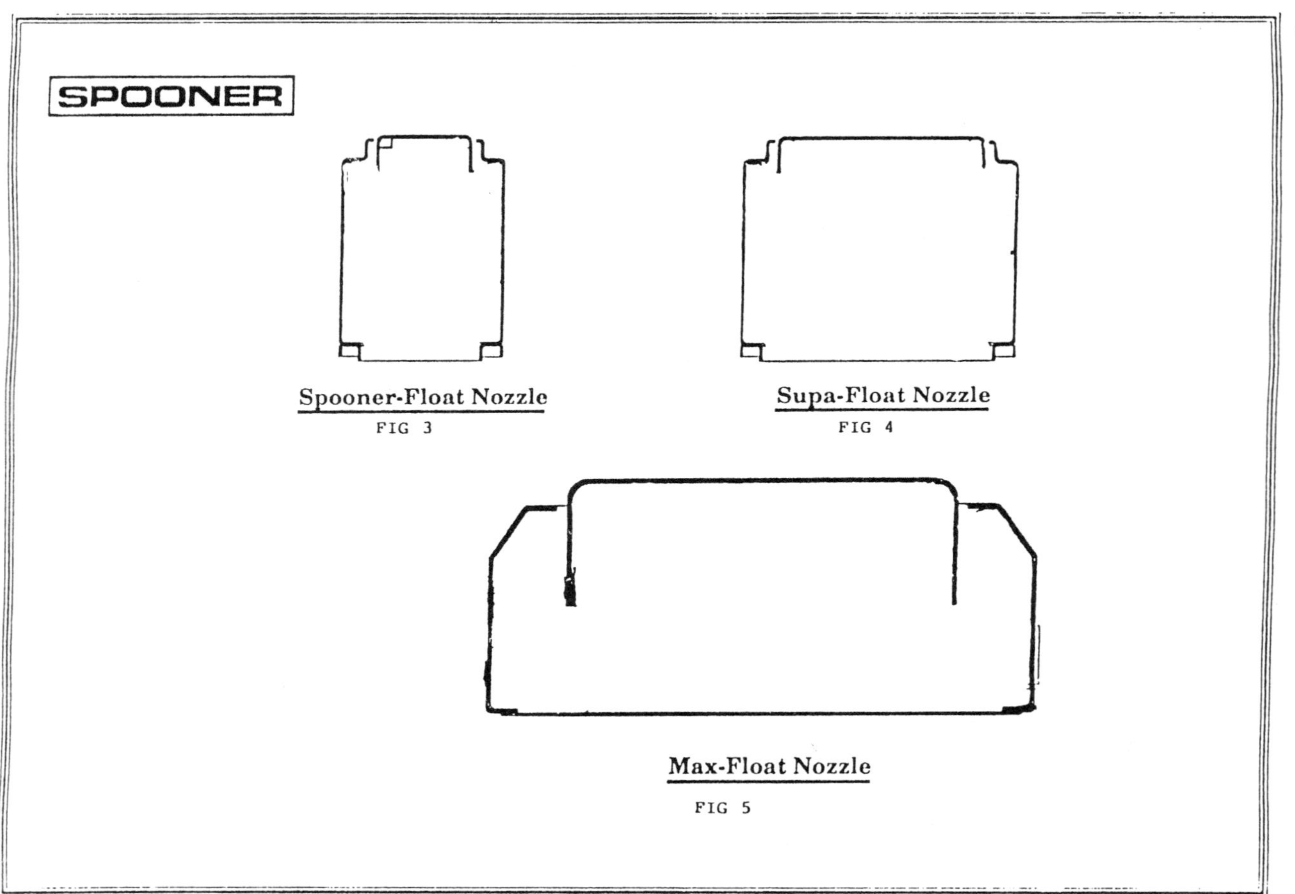
SPOONER
Spooner-Float Nozzle
FIG 3
Supa-Float Nozzle
FIG 4
Max-Float Nozzle
FIG 5

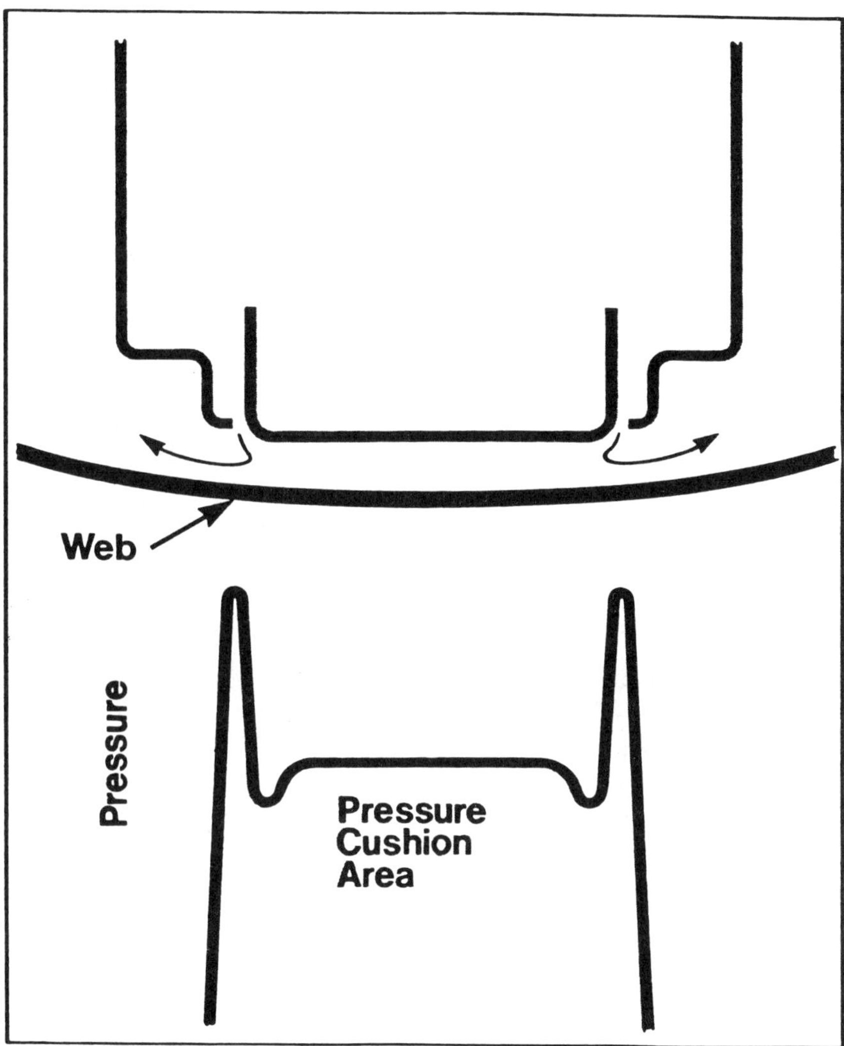

Fig 6

for, although each portion of the geometry can be optimised in isolation, it is the performance of the whole which is paramount.

From Fig. 6 (upper) it can be seen that the air flow exiting the two parallel slots impinges upon the web in such a manner as to generate a pressure profile as on Fig. 6 (lower). It is this generated pressure profile that is the key to successful performance. It is essential that this profile is perfectly symmetrical. This is guaranteed by very close control of the manufacture of the nozzle.

The pressure profile is made up of two distinct regions, namely

1. The impingement peaks

2. The supporting cushion.

During operation the impinging air scours away the boundary layer at the web surface and, because of the heat transfer, drying takes place. Fig. 7

At the same time air exiting from the nozzle slots via the action of the coanda effect travels around the radius top corner and forms the very stable cushion upon which the web is conveyed. The forces generated in the air flotation cushion are high in magnitude and are used to offset the force generated by the tensions in the web. Because the nozzles are alternately pitched at the top and bottom the effect of this is to generate a sinusoidal wave form which further enhances the web stability. (Fig. 7).

4 CHARACTERISTICS OF A GOOD AIR FLOAT DRYER

1. Must support the web being processed in a true defined contactless mode and hence eliminate the possibility of product contamination.

2. Must achieve complete web stability.

3. Must handle a range of webs.

4. Performance of the dryer must not be compromised by machine tension fluctuations.

5. Must achieve high uniform heat transfer rates.

The web being processed should pass through the drying system without any tendency to flutter, buzz or vibrate and this should be achieved under all operating conditions, irrespective of the air velocity being employed, whether it be extremely high or low.

Spooner - Sine wave configuration
Exhaust
Area
Boundary
Layer
Impingement
Areas
Coating
Substrate

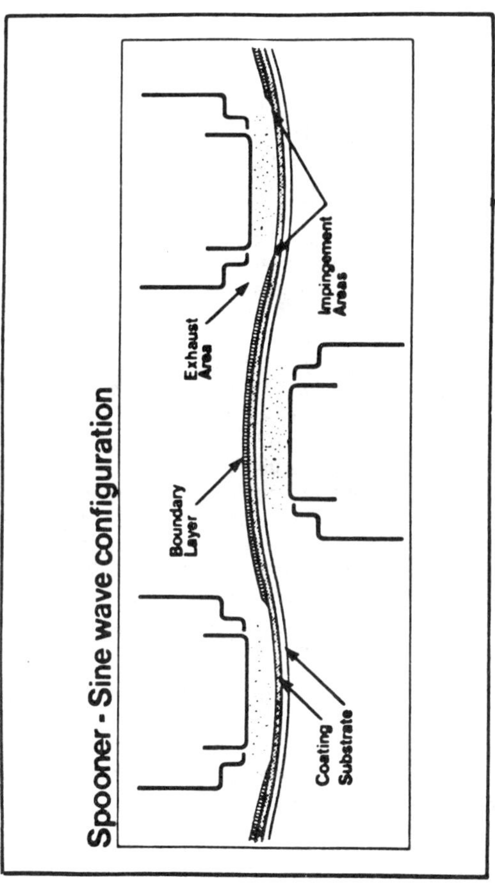

Fig 7

Operating velocities in the range 10 - 100m/second are achievable without deterioration with pressure differences in the upper and lower air bars or vice versa capable of being as high as 2:1.

Allied to this web stability the system should act upon the web by virtue of the alternate pitching of the upper and lower air bars in such a way as to maximise the inherent stiffness in the product. This alternating of air bars generates a sinusoidal wave form which imparts a beam stiffness because of the corrugating effect across the product, which in turn suppresses any latent tendency for the product to wrinkle, flute or curl.

Coupled with this stability the ideal system operates at high web to air bar surface clearance which, although variable, is a constant during process operations i.e. there is no tendency for the clearance to change and permit the risk of touch down. This operating clearance must not be adversely affected by the adjustment of impingement air velocity on one or both sides of the web. Clearances in the range 6 - 16 mm are easily achievable.

The good system displays a wide operating range of heat transfer coefficients on both sides of the web, which are not adversely affected by operation within the web to air bar clearance range of 6 - 16 mm, high heat transfer and, hence, high evaporation whilst at the same time retaining the capability of operating with low heat transfer rates and hence low evaporation rates. To understand how this is achieved the correlation between Nusselt, Reynolds and Prandtl numbers embodied in the convection equation needs addressing.

$$Nu = A\,Re^{n}\;Pr^{0.33}$$

$$\text{where } Nu = \frac{hd}{K} \qquad Re = \frac{pVd}{N} \qquad Pr = \frac{CpN}{K}$$

When under examination it transpires that the heat transfer achievable is governed by the gas velocity and the free area of the system. Many years of experience in the field has led to the determination of the most economic value of free area of the nozzle system. It therefore follows that the Reynolds exponent of the system is the governing factor with respect to heat transfer via the nozzle velocity, and it is this value that Spooner have sought to maximise in an effort to achieve the maximum performance out of the system for a given power input.

The correctly designed air float system does not dictate the operating web tension level. The product

PRESSURE PROFILE ON 9.1m (360 in.) LONG AIR BAR
Pressure in Air Ba
152 (6.0)
146 (5.75)
140 (5.5)
136 (5.37)
133 (5.25)
127 (5.0)
mm. W.G. (in. W.G.)
9.1 metres
Average Pressure
1.8% Low Variance
1.3% Top Variance
Air Inlet Positions

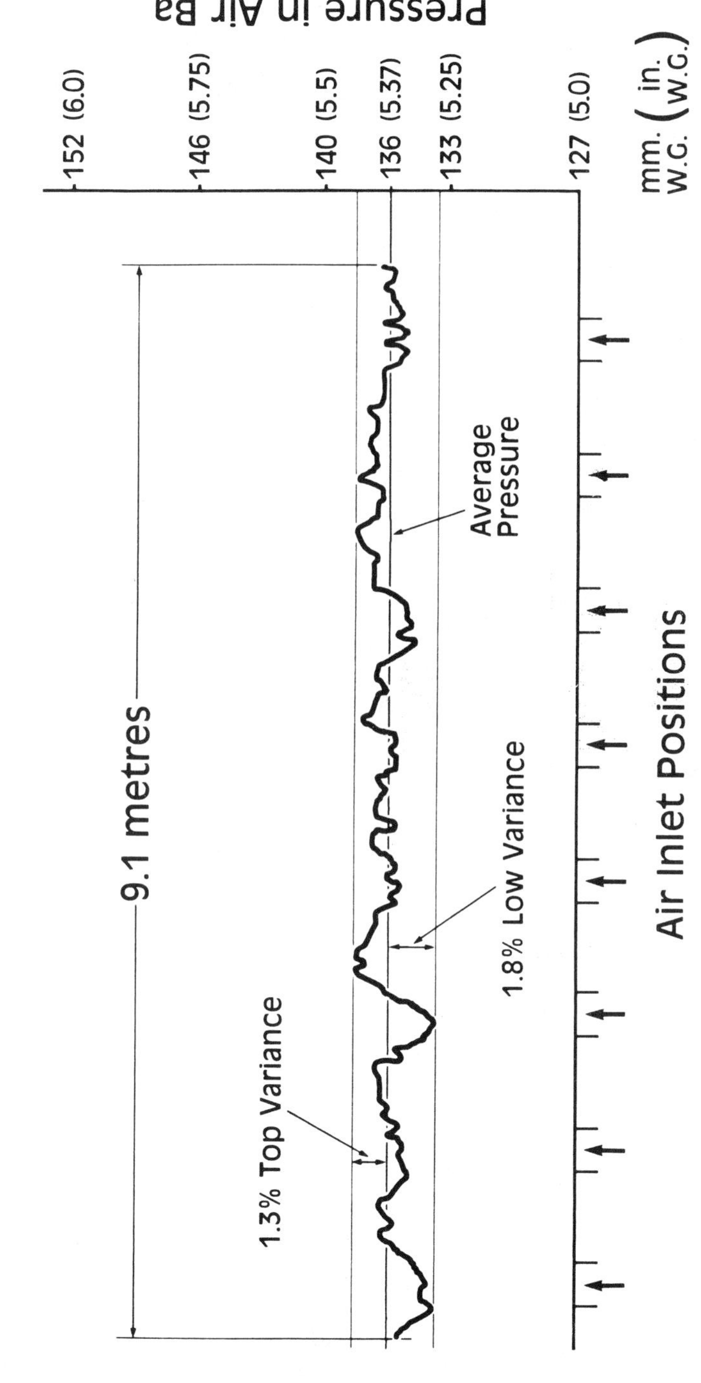

Fig. 8

and the process requirements are the deciding factors used to determine the tension at which the system operates. The Spooner Float system operates in the range 1-120 kg/metre of product width, dependent on product type, without any deterioration in the operating characteristics.

Collectively all of the above conditions should be present irrespective of the process being carried out. These characteristics are obtained all in the one design of the Spooner Float State of the Art air bar. Air flotation lends itself to the manufacture of equipment that will achieve very uniform cross-machine drying profiles (Fig. 8) even on very wide process web widths. Spooner have experience of supplying very wide machines to the industry of circa 9 metres but, by good air circuit design, such machinery can hold very close tolerances of temperatures, velocity and hence evaporation rate. Such is the expertise of Spooner in this field that process guarantees are given with confidence but, more importantly, they are obtained in practice.

In summary Spooner services the market place via very basic sound principles.

i) Continued Research and Development to enhance the very sound technological foundations laid down many years ago.

ii) A wealth of unequalled experience in the field.

iii) A superbly equipped Test Centre in which existing and new processes can be tried and proven.

iv) A very sophisticated computer drying model to accurately simulate the process under scrutiny.

In this way the market place is assured of Spooner supplying the technical correct solution to the demands on air flotation dryers.

5 AIR TURN ROLL

Introduction

Prior to the development of air flotation turning technology double sided coated webs were turned by the use of conventional roll turning systems such as chill rolls. Whilst adequate for some applications, problems arose, particularly when processing with difficult coatings of low viscosity and solids content and/or high coating weights. This led to product quality reduction for a number of reasons, namely:

1. Intimate contact with the web turning apparatus.

2. "Picking" problems on the surface of the drying cylinders.

The Air Turn Roll was conceived as a method of overcoming these problems.

A number of "contactless" turning systems had previously been tried, consisting mainly of perforated rolls. However none of these had been wholly successful.

Two very simple goals were set for the development of the air turn.

1. It must be able to maintain a good working clearance between the web and the unit itself through a pre-determined wide operating tension range.

2. It must demonstrate the ability to turn the web through any angle whilst maintaining sheet stability without forming creases.

As a starting point it was decided that there was absolutely no point in attempting to "re-invent the wheel" so to speak and as such the basis for the design was to be a very slightly modified Spooner Float air bar, complete with its remarkable web handling characteristics.

Early prototypes consisted of air bars mounted directly onto a cylindrical pressure chamber. This in itself was successful as an air flotation system, allowing the spent air to move into the space between the nozzles before moving sideways at low velocity. However Spooner suspected that a performance enhancement could be gained with further development work.

The initial major development was the introduction of plates between the air bars with air dams towards the edges of the floating web.

The purpose of this was to maximise the cushion area presented to the web to oppose the prevailing tension.

Fig. 9

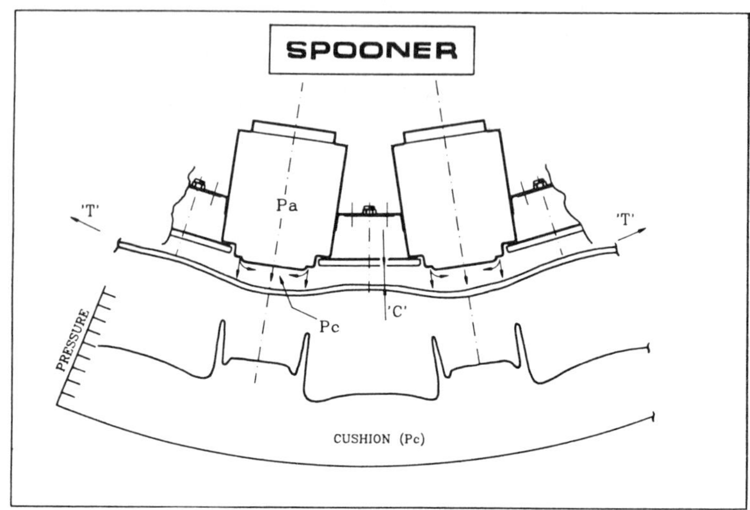

This reduction in available cross-sectional area for exiting air flow yielded the following benefits:

1. Enhanced web stability with a more uniform clearance over the full web width.

2. Because the velocity of the exiting air was increased it imparts an increased viscous drag on the web equal and opposite in direction from the centre line of the web outwards, which has the tendency to inhibit the formation of creases.

3. Due to this reduction in air escape area there is a considerable reduction in the volumetric flow requirements of the fan system.

The addition of the edge dams, apart from minimising the air escape area, also formed a series of air locks over the deckle width. The action of these air locks was to gradually reduce the cushion static pressure between the unit and the web, such that the web edge exhibited no tendency to flutter or buzz, and enabled the air turn unit to process pre-determined deckle widths without a loss in handling performance.

Flotation Height of the Web

Following extensive trial work it was decided that a clearance of 6 mm between the air turn and the web was to be used as an optimum, this being a good working clearance ensuring trouble-free operation without the possibility of contamination of the surface during operation, whilst also keeping the nozzle velocity requirement to a minimum, thus further minimising fan power usage.

Fig. 10

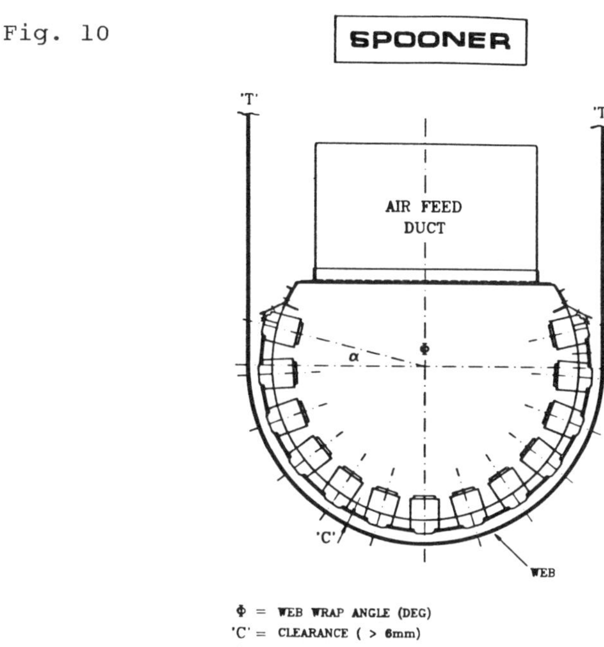

Tension

The 6 mm clearance is now a standard feature of the air turn system, irrespective of web tension. To add a further degree of flexibility to the system the air velocity can be adjusted in order to maintain the 6 mm clearance regardless of the tension up to the design maximum, normally via a remotely operated damper.

Fig. 11

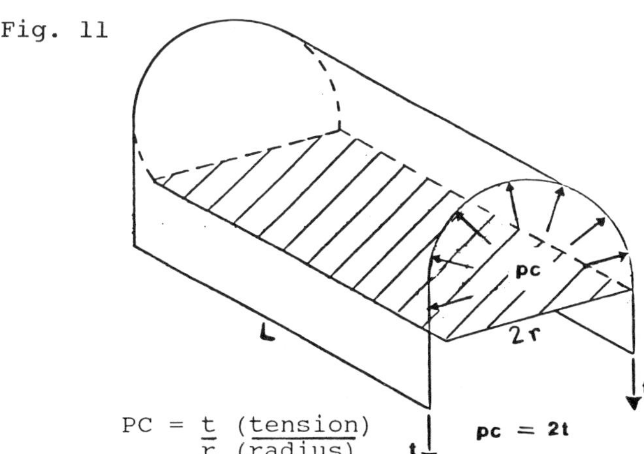

Whilst this is adequate for machines where the machine tension remains constant, there are machines that are subject to tension variation or differing tensions are required for differing operations. To facilitate the control of such a situation an automatic clearance control system has been developed.

Fig. 12

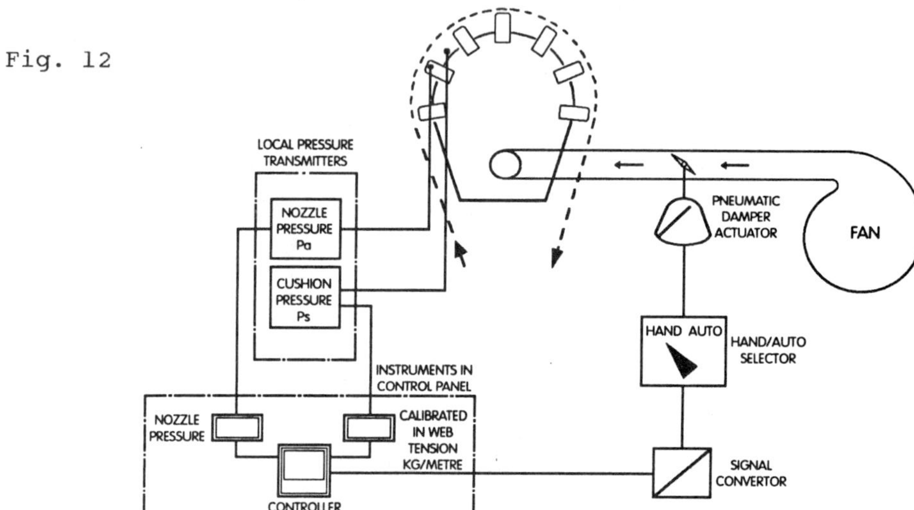

This is based on the principle that for flotation to take place the tension forces exerted on the unit must be opposed by the cushion pressure developed by the unit. Any change in tension will clearly affect the cushion pressure and hence, by constant monitoring of the cushion pressure, the air flow can be automatically regulated to compensate for any tension fluctuations. It also thus measures the prevailing machine tension accurately and non-contactly which can be displayed if so desired.

Further Developments

As was mentioned earlier the primary reason for the development of the air turn was to overcome web handling problems. However in some instances it became apparent that the application of an air turn roll can result in increased machine speed. This can be attributed to the impingement effect of the nozzles giving rise to forced ventilation on the wet sheet. It was found that this form of assisted drying improved with elevated web temperature at the entrance to the air turn unit.

Following on from this it was logical to add some degree of heat into the impingement air system to assist with the raising of the solids content on the setting of a coating before it comes into contact with the next contact apparatus.

The addition of this heat is clearly feasible by two different systems.

1. The simplest way is to use low grade heat such as turbine pass-out steam that could be put through a heat exchanger before entering the supply system. This system is of course, non re-circulatory and can only be considered where the low grade heat could be put to no better use.

2. Should it be considered beneficial to use higher temperatures than this then it is essential to incorporate a recirculatory system to maximise the thermal efficiency of the unit.

Fig. 13

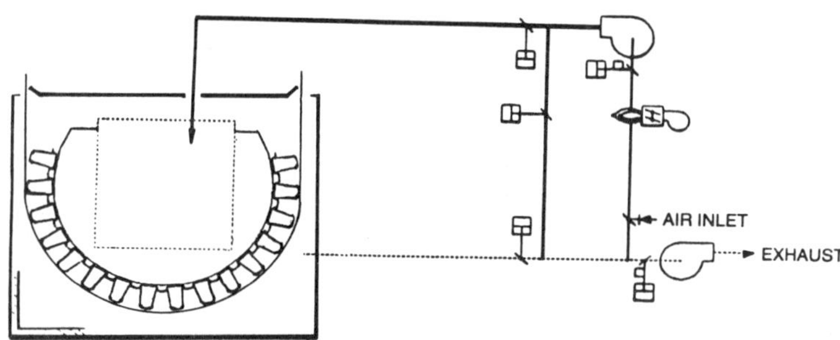

The first unit of this design was installed on a paper machine in Northern Germany where the air turn followed a double-sided coater and a gas fired infra-red system. In-house computer simulations showed that it would be possible to either raise sheet temperature or increase evaporation by controlling the humidity of the recirculation air.

In practice the above was found to be true and now the mill has the ability to control the drying regime depending on the grade of paper being processed.

Crease Control

Further development of the basic air turn concept has been to improve the crease removal characteristics, this is particularly important when the unit is operated on conjunction with the new generation bar metering size presses where experience has shown that the speed and nip configuration can result in severe crease generation.

Following development work undertaken at Spooner a world-wide patent was taken out on a device aimed at combatting this phenomenum (Fig. 14).

Essentially the air turn unit is bowed in the sheet width direction or has the additional feature which enables the internal pressure to be varied across its width such that, although the cushion pressure is a constant, when a crease is exhibited there exists the possibility of operating that specific portion of the air turn at an increased flotation height, thus creating an extra requirement for web surface and hence removing the crease. The first unit of this kind was recently installed in Germany and although in-house testing looked very promising we are awaiting initial performance testing results before pronouncing on the degree of success of the unit.

Fig. 14

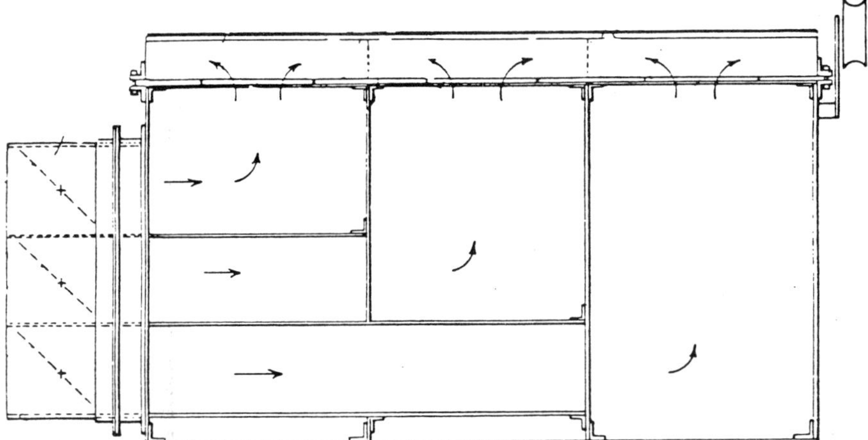

However, the development work does not end there for it is hoped that in the very near future there will be another three products to add to the air turn family, thus ensuring new and interesting markets for Spooner as well as satisfying the market place demand for ever-increasing production rates, product quality and product sophistication.

Possible Air Turn Applications

Typical examples of applications of the Spooner Air Turn:

1. Following Size Press (Fig. 15)

2. Following Coating Equipment

3. Between the various levels of a multi-pass air flotation dryer

to name but a few.

In Summary

The Spooner air turn roll is a unique system offering the following.

1. Non-Contact Web Handling.

2. Extensive tensions range of primarily 10 to 100 kg/metre width of product (however units can be designed to cater for tensions far in excess of this, if so desired.)

3. Web turning angle 20 - 180 degrees.

4. Unlimited web width.

5. Cold or heated operation.

6. Automatic control of clearance height with tension display.

7. Minimum diameter of 500 mm with no real limit on the upper size.

Fig. 15

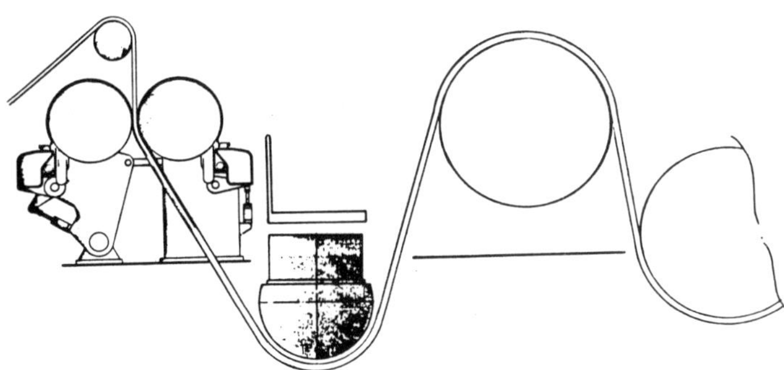

Testimony to the performance of the air turn unit is the customer reference list which now stands at approximately 100 units.

6 JET FOIL TM

Conventional steam heated dryers have been used in the paper making process for over 160 years for very good reasons.

i.e.
- high heat and mass transfer rates
- sheet flattening (ironing effect)
- draw control
- lateral stability (intimate contact guiding effect)
- surface smoothness

However, the Jet Foil$_{TM}$ was conceived as an alternative technology designed to be used in the existing paper machine configuration to overcome the disadvantages often attributed by the paper maker to problems inherent in the application and design of steam heated drying cylinders, such as:-

- sheet picking
- cross machine profile
- drying rate limitation
- crease formation
- surface contamination
- speed and draw variation
- thermal inertia

Development

Having agreed that a self-contained air flotation cylindrical dryer was feasible as a method of overcoming the problems outlined above, it was clear to Spooner that the Jet Foil$_{TM}$ would need to conform to special criteria.

1. It must be able to achieve a good working clearance between the web and the Jet Foil$_{TM}$ surface, whilst maintaining a stable web condition over the full sheet width.

2. It must be able to achieve higher drying rates than those already achievable with steam cylinders.

3. The unit must feature a self-contained air recirculation system in order to minimise fuel usage.

4. The Jet Foil$_{TM}$ cylinder must have similar geometry and be able to fit into the space vacated by the steam cylinder.

5. It must be totally compatible with the paper machine in terms of threading and general operation.

Advantages

In meeting these conceptual design requirements the advantages offered by the Jet Foil$_{TM}$ system can be summarised as follows:

- contactless operation
- higher evaporation rates
- low thermal inertia
- no drive requirement
- wide operating tension range
- free shrinkage
- accurate CD profiles
- fast control response

Jet Foil$_{TM}$ Air Support System

The basis of this patented system is a unique, specially designed Jet Foil$_{TM}$ air bar.

Fig. 16

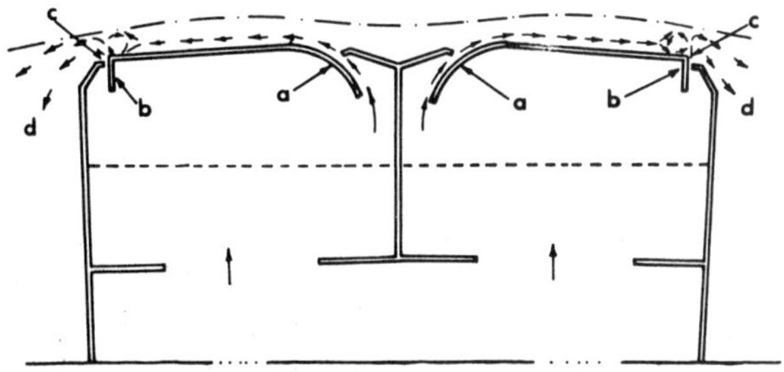

This air bar design features two opposed air foil nozzles that utilise "coanda" effects to give precise sheet positioning and extremely stable flotation. The opposed air streams pass over extended surfaces in order to maximise the area of cushion support.

A second modified slot ensures that the optimum flotation height is maintained in the vicinity of the sub-atmospheric pressure region generated in the return air spaces between the air bar units.

In the Jet Foil$_{TM}$ cylinder the impingement jets are designed to promote regions of micro turbulence at the sheet surface to generate high heat transfer rates which in turn, through correct machine design and operation, provide the driving force for effective water removal.

On leaving the impingement site the "spent" air continues to move in a radial manner and is taken immediately into the sub-atmospheric region between the air bars and into the re-circulation system.

Great emphasis is placed on ensuring that the extraction of spent air is done in such a way as not to hinder the evaporation mechanism or create any unevenness.

The combination of both high and low pressure areas not only supports the web in a very stable manner, but also imparts a gentle sine wave formation into the web as it passes around the cylinder. This has the effect of helping to suppress the formation of creases and flutes and also any tendency for curl at the edges of the web.

Fig. 17

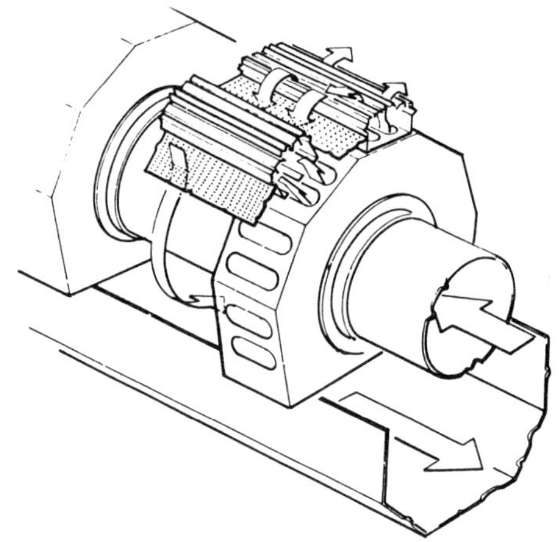

Arrangement Of The Jet Foil$_{TM}$

As the Jet Foil$_{TM}$ cylinder was designed to directly replace the conventional steam cylinder, the standard diameter is 1,500 mm. However, if the process requirement dictates a larger diameter and the physical constraints on positioning allows it, it can be manufactured in excess of 3 metres diameter, since it operates at relatively low pressure and as such is not subject to the constraints imposed by the pressure vessel requirements of the conventional steam cylinder.

The air bar units are arranged in a circular form to embrace the required sheet wrap angle, this is normally 230° involving the use of 7 air bar units.

In most situations the Jet Foil$_{TM}$ cyinders would normally be installed in multiples of two to ensure even treatment to both sides of the web. However, this is not essential, odd numbers of cylinders can be employed effectively without adversely affecting the process or product.

Fig. 18

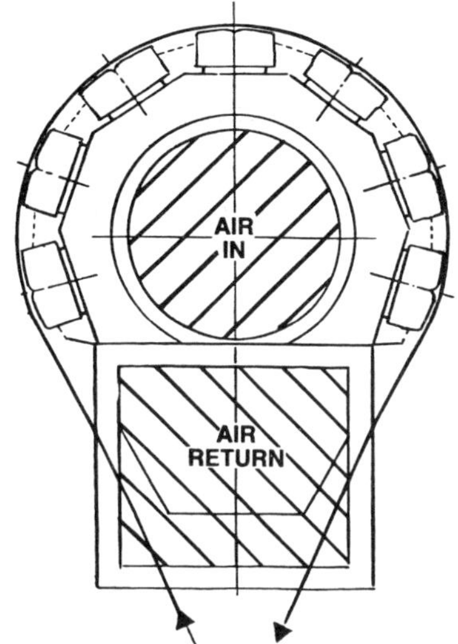

Arrangement of the Air System

Typically one pair of Jet Foil$_{TM}$ cylinders would be serviced by one air supply system positioned at the drive side of the machine.

Fig. 19

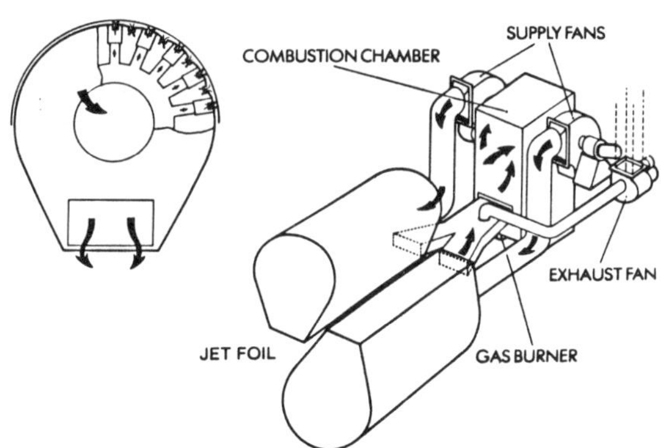

Briefly the heated air is delivered to each Jet Foil$_{TM}$ cylinder via its own supply fan. Having impinged onto the sheet, the air is returned to the fan and heater units for recirculation. To maintain optimum humidity within the system a pre-determined amount of exhaust air is removed from the system and a corresponding quantity of fresh air introduced into the system. This can be varied right through the range to give zero to 100% recirculation if so desired. All of this recirculation takes place away from the cylinder itself, there is no requirement for air to be taken in or lost at the surface of the Jet Foil$_{TM}$ cylinder.

Operating Temperature

The Jet Foil$_{TM}$ is capable of being heated by almost any available heat source and this is the factor that governs the temperature capability of the system.

By direct gas firing, impingement air temperatures in excess of 300°C are achievable.

Operating Performance

Considering the temperature potential increase and the improvement to heat transfer coefficient in general, the evaporation potential of the Jet Foil$_{TM}$ cylinder lies in the range of 1.5 to 2 times that realisable from a conventional system.

This comparison assumes that the maximum available temperature of the conventional cylinder can be utilised in practice. In truth this is often limited by surface picking or contamination problems such that the potential difference is even higher. Such a position would be the first after drying cylinder following a size press or coater.

It is worth noting that the prototype installation at Billingsfors paper mill operating with 2 Jet Foil$_{TM}$ cylinders 3.3 metres wide, following a size press, achieved an average evaporation of 25 kg/m^2/hr over the two Jet Foils$_{TM}$ when being heated by steam. The previous evaporation rate from the two conventional cylinders was in the range 5 to 10 kg/m^2/hr.

Applications For Jet Foil

By way of example there are a number of positions on the paper machine where the Jet Foil$_{TM}$ system may be utilised to good advantage:-

- After size press or two sided coating (Fig. 20)
- Drying limited pre or after dryer sections
- Prevention of picking after presses
- Free shrinkage of extensible papers (sack Kraft)

Operating Parameters

Operating width	1,000 - 6,000 mm
Diameter	1,000 - 3,000 mm
Sheet tension	2 - 30 kgs/metre
Operating temperature	20 - 350°C
Evaporation potential	10 - 40 kg/m^2/hr
Maximum machine speed	Over 1,000 m/min

Fig. 20

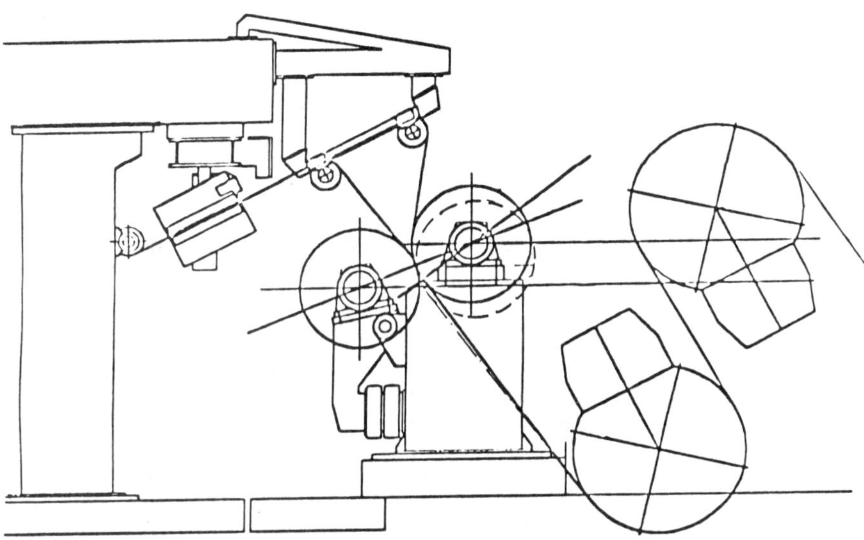

Conclusion

I hope this paper has shed some light on the technical complexities of air float drying and non-contact web handling but it is a challenge that Spooner relish for it is only through continued well structured R & D work that today's innovative concepts become the solution to tomorrow's ever-technically spiralling process problems.

Environmental Aspects of Solvent Drying of Coatings and The Case for Catalytic Incineration

John C. Ashworth

ADVANCED SOLIDS PROCESSING CONCEPTS (ASPCETS), 'THE KNOLL', OLDWICH LANE EAST, FEN END, NEAR KENILWORTH, WARWICKSHIRE CV8 1NR, UK

ABSTRACT

Due to recent legislation, operation of solvent-based dryers for coating applications will be subject to much tighter environmental controls. This is examined with respect to the properties of solvents such as LEL values, autoignition temperatures, maximum solvent loading (MSL), exhaust emission limits (EEL) as specified in the HMIP Process Guidance Notes, and the pollution control index (PCI) based upon the ratio of EEL to MSL. For common VOC's it is shown that very high solvent recovery efficiencies of the order of 99.9% are needed when processing effluent airstreams.

Various techniques of solvent recovery and treatment are compared such as low temperature condensation, fixed-bed adsorption, liquid absorption, thermal incineration and catalytic incineration. For applications such as drying of coatings it is shown that catalytic incineration will frequently prove most cost effective because the heat of combustion of the solvent can be utilised to provide an energy self-sufficient system.

The main features of a new catalytic incinerator, developed by the Italian Company, Eco Trend System, are described. This equipment has been assembled together to provide compact packaged units suitable for easy connection to existing installations. These are of modular construction with capacity to treat 2500 Nm^3 per hour. The units utilise an unique two-stage catalytic combustion system to give optimal thermal conversions and solvent removal efficiencies from the airstream.

The design of economic solvent treatment systems is examined via a case study for coating of a plastic film with an ethyl acetate based adhesive. This shows the VOC concentration in the airstream dramatically affects the volume of air to be treated and size and cost of equipment required. A comparative study also shows that catalytic incineration equipment is typically about 70% of the cost of thermal incineration plant of equivalent capacity. But the best overall performance is achieved with a combined solvent recovery/catalytic incineration system. For the latter case, the value of the recovered solvent enables the pollution control project to become self-financing with paybacks of 1 to 2 years.

The key to effective solvent handling is the maximisation of the VOC content in the exhaust air. Therefore the dryer system should be carefully configured to approach the 50% LEL limit as closely as possible. To achieve such results, detailed understanding of the performance of the drying system from zone to zone must be available such as can be provided by modelling software techniques developed by the author. These procedures enable cost effective VOC treatment systems to be developed for most industrial applications. The modular nature of the catalytic combustion units also offers flexible modes of operation needed when plant is used for multiproduct manufacture.

INTRODUCTION

As a result of recent legislation, Environmental Protection Act 1990, all industrial processes will be subject to strict licensing and high standards of pollution control. All sectors of industry must implement Integrated Pollution Control (IPC) procedures within the next several years. For the paper manufacturing and coating industries the timescale for upgrading to the full standard is scheduled to be completed during 1996 to 1998 for existing operations. However new processes, and existing processes which have been the subject of substantial change since 1st April 1991 are immediately subject to the IPC regulations (ref.1).

For the coating industry, IPC will be mainly concerned with the control of releases of organic solvents to the environment subject to:

BATNEEC - the best available techniques not entailing excessive cost are used to prevent or, if that is not practicable, to minimise the release, and to render harmless any prescribed substances which are released into any environmental medium;

BPEO - when a process is likely to involve releases into more than one medium (air, water or land), the best practicable environmental option is achieved so as to have the least effect on the environment as a whole.

This paper sets out to provide guidance as to the technologies available to handle emissions arising from the drying of solvent based coatings, and to identify plant configurations which best meet the constraints of integrated pollution control procedures.

PROPERTIES OF SOLVENTS

The problems of pollution control are closely linked to the health and safety aspects of handling solvents. The key properties of 36 solvents of most common use in the converting industry are presented in Tables 1 and 2. The significance of the various properties are as follows:

LOWER EXPLOSION LIMIT (LEL) - this defines the lowest concentration of solvent vapour in air, which if ignited will just propagate flame. When evaporating solvents in air, the LEL value effectively specifies the maximum safe working concentration of solvent in the dryer.
In practice, a significant factor of safety is adopted such that the maximum concentration permitted is only 25% LEL, unless suitable instrumentation to monitor solvent concentrations is installed on the dryer. For the latter situation operation at 50% LEL is feasible.

AUTOIGNITION TEMPERATURE - this defines the minimum temperature for a solvent to initiate self-combustion in air in the absence of a spark or flame. Hence for safe operation air temperatures used within the dryer should be kept below this limit. A factor of safety of 50 to 100C is usually recommended.

MAXIMUM SOLVENT LOADING (MSL) - this defines the maximum practical solvent loading in the airstream on a mass ratio basis where operation at 50% LEL is assumed. For most solvents, large quantities of air are required for safe handling which in turn leads to high energy loads.

DEWPOINT TEMPERATURE AT MAXIMUM SOLVENT LOADING - this defines the temperature to which the air/solvent mixture must be cooled before condensation is initiated.

EXHAUST EMISSION LIMIT (EEL)- this defines the maximum loading in the air stream exhausted from the process to meet pollution control standards and emission limits. These figures have been defined in the series of Process Guidance Notes produced by Her Majestys Inspectorate of Pollution (refs. 2 to 6). These Guidance Notes set emission limits for the converting industry for common VOCs at 50 mg per normal metre cubed, and for chlorinated VOC's at 20 mg per normal metre cubed.

TABLE 1

Safety Aspects of Handling Organic Solvents (MSL Values)

Solvent	Formula	Mole Wgt /mol g^{-1}	Lower Explos Limit /%	Auto Ignit Temp /C	Max Solv Load /gg^{-1}
dichloromethane	CH2C12	84.93	1.55	662	.0229
methanol	CH4O	32.04	7.30	464	.0419
methyl chloroform	C2H3C13	133.41	8.00	537	.1919
vinyl trichoride	C2H3C13	133.41	8.60	460	.2069
ethanol	C2H6O	46.07	4.30	423	.0349
ethylene glycol	C2H6O2	62.07	3.20	413	.0348
ethylenediamine	C2H8N2	60.10	2.70	385	.0284
acetone	C3H6O	58.08	2.60	538	.0264
methyl acetate	C3H6O2	74.08	3.10	472	.0403
dimethyl formamide	C3H7NO	73.10	2.20	445	.0281
2-nitropropane	C3H7NO2	89.10	2.50	425	.0389
isopropanol	C3H8O	60.10	2.00	399	.0210
n-propanol	C3H8O	60.10	2.00	371	.0210
methyl cellosolve	C3H8O2	76.10	1.80	285	.0239
vinyl acetate	C4H6O2	86.09	2.60	427	.0391
methyl ethyl ketone	C4H8O	72.11	1.80	516	.0226
tetrahydrofuran	C4H8O	72.11	2.00	321	.0251
diethylene oxide	C4H8O2	88.11	2.00	180	.0307
ethyl acetate	C4H8O2	88.11	2.20	427	.0338
dimethyl acetamide	C4H9NO	87.12	2.00		.0304
butanol	C4H10O	74.12	1.40	343	.0180
cellosolve	C4H10O2	90.12	1.80	235	.0283
n-butylamine	C4H11N	73.14	1.70	312	.0216
propyl acetate	C5H10O2	102.13	1.80	460	.0320
cyclohexanone	C6H10O	98.14	1.00	420	.0170
cyclohexane	C6H12	84.16	1.33	260	.0195
isobutyl methyl ketone	C6H12O	100.16	1.40	459	.0244
n-butyl acetate	C6H12O2	116.16	1.70	421	.0344
isobutyl acetate	C6H12O2	116.16	2.40	423	.0487
diacetone alcohol	C6H12O2	116.16	1.80	603	.0364
cellosolve acetate	C6H12O3	132.16	1.70	380	.0391
hexane	C6H14	86.18	1.20	234	.0180
butyl cellosolve	C6H14O2	118.18	1.10	244	.0226
toluene	C7H8	92.14	1.20	536	.0192
n-heptane	C7H16	100.20	1.20	223	.0209
xylene	C8H10	106.17	1.00	464	.0184

TABLE 2

Safety Aspects of Handling Organic Solvents (EEL Values)

Solvent	Exht Emiss Limit /mgl^{-1}	Exht Emiss Limit /ppm	Dewpt Temp MSL /C	Dewpt Temp EEL /C	Pollut Contr Index /%
dichloromethane	20.0	5.55	-5.0	-111.7	.0716
methanol	50.0	36.80	-1.0	-74.7	.1008
methyl chloroform	20.0	3.53	-3.0	-99.6	.0088
vinyl trichoride	20.0	3.53	31.6	-80.0	.0082
ethanol	50.0	25.58	4.4	-66.9	.1190
ethylene glycol	50.0	18.99	94.8	6.3	.1187
ethylenediamine	50.0	19.61	19.8	-50.2	.1453
acetone	50.0	20.29	-32.1	-96.4	.1561
methyl acetate	50.0	15.91	-26.9	-94.3	.1027
dimethyl formamide	50.0	16.13	36.8	-38.0	.1466
2-nitropropane	50.0	13.23	11.9	-78.4	.1059
isopropanol	50.0	19.61	-1.3	-62.1	.1961
n-propanol	50.0	19.61	10.4	-53.4	.1961
methyl cellosolve	50.0	15.49	21.4	-42.0	.1721
vinyl acetate	50.0	13.69	-18.8	-90.2	.1053
methyl ethyl ketone	50.0	16.35	-21.4	-85.5	.1816
tetrahydrofuran	50.0	16.35	-29.6	-92.5	.1635
diethylene oxide	50.0	13.38	-3.7	-77.5	.1338
ethyl acetate	50.0	13.38	-16.4	-83.4	.1216
dimethyl acetamide	50.0	13.53	54.0	-30.2	.1353
butanol	50.0	12.99	21.0	-41.7	.1856
cellosolve	50.0	13.08	29.3	-35.6	.1453
n-butylamine	50.0	16.12	-20.3	-84.1	.1896
propyl acetate	50.0	11.54	-1.1	-69.9	.1282
cyclohexanone	50.0	12.01	22.9	-46.8	.1201
cyclohexane	50.0	14.01	-27.2	-88.0	.2106
isobutyl methyl ketone	50.0	11.77	3.8	-61.5	.1681
n-butyl acetate	50.0	10.15	15.6	-59.1	.1450
isobutyl acetate	50.0	10.15	12.3	-65.9	.0846
diacetone alcohol	50.0	10.15	48.3	-40.0	.1128
cellosolve acetate	50.0	8.92	41.5	-38.4	.1049
hexane	50.0	13.68	-35.7	-91.8	.2279
butyl cellosolve	50.0	9.98	48.8	-17.9	.1814
toluene	50.0	12.79	-6.1	-71.8	.2132
n-heptane	50.0	11.76	-13.7	-75.6	.1961
xylene	50.0	11.10	15.9	-53.3	.2220

It should be noted that these emission limits are very strict and for most common solvents are much lower than those needed to meet the time-weighted average 8 hour long-term exposure limits for health requirements set out in the COSHH regulations (ref. 7).

Another concern for reasons of international competiveness is that these limits are much tougher than those required by authorities in other countries at present. However one of the reasons for the strict policy adopted by HMIP is to anticipate the much tighter air pollution standards agreed by the EEC to come into force in several years time. Nevertheless it is interesting to compare standards adopted elsewhere. The U.S.A. Environmental Protection Agency (ref. 8) recommends emissions to the ecology for some common organics such as propane may be 100 times higher than those permitted for health purposes. By contrast, for substances which are carcinogenic such as benzene, the ecological standards adopted are more strict than those required for health purposes.

An alternative viewpoint is provided by the German TA Luft specifications. In this instance each solvent is categorised according to class 1, 2 or 3 with emission limits of 20, 100 and 150 mg per normal metre cubed respectively. This means the standard for the most toxic class 3 solvents are similar to the figures presented in Table 2, but the German standard is more relaxed for the least toxic substances such as acetone and ethanol.

DEWPOINT TEMPERATURE AT EXHAUST EMISSION LIMIT - this defines the temperature to which the solvent contaminated airstream must be cooled to condense out the solvent to reach the exhaust emission limit.

POLLUTION CONTROL INDEX (PCI) - this is defined as the percentage ratio of the exhaust emission limit to the maximum solvent loading permissible. In effect, the higher the value of this index the easier it is to process the exhaust stream from the dryer. Note also that the percent solvent recovery required is equal to (100 - PCI).

METHODS OF SOLVENT RECOVERY AND REMOVAL FROM AIRSTREAMS

Several options are available for treatment of solvent contaminated airstreams. Each technique has its advantages and disadvantages for individual cases as discussed below.

1. LOW TEMPERATURE CONDENSATION - uses a coolant or refrigerant fluid via a heat exchanger to chill the airstream to condense out the solvents. Although this technique is straightforward, it is generally most appropriate when applied to small flows, and when high concentrations of solvent are present such as may be possible via inert gas operation. This is illustrated by Table 2 which shows very low condenser temperatures, less than -50C, are required by most solvents to enable emission standards to be met. It should also be noted that significant quantities of water vapour are likely to be present in the air which will also condense out during chilling. Further processing will be required to purify the solvent.

2. FIXED BED ADSORPTION - uses packed bed of granules such as activated carbon, molecular sieves or silica gel to desorb the solvent from the airstream. This is one of the most popular methods for removing solvents present in low concentrations and large volume flows.

   A practical system usually comprises two or more beds in parallel, such that one bed is desorbing solvent while the second bed is being regenerated with steam. The solvent stripped off by the steam is condensed. Further purification may be required by rectification for water-soluble solvents.

   The disadvantages of fixed bed adsorption are that the performance of the bed falls off sharply with temperature so exhaust gases from dryers will need to be cooled prior to treatment. Additional energy is also required for regeneration. For example, in their brochures for activated carbon systems Lurgi note steam consumption figures of between 2 and 4 kg per kg solvent and another 1 to 1.5 kg per kg solvent when rectification is required.

3. LIQUID ABSORPTION - use of circulating liquid, usually by countercurrent contact between the gas and liquid streams in a packed or tray column, to absorb solvent from airstream. The success of this technique depends on whether or not a suitable absorption liquid is available which has a low volatility and does not form an azeotrope with the solvent to be recovered. For example, water can be used to extract miscible solvents such as acetone. At intervals the absorbent liquid is processed to recover the solvents by distillation, and recycled back to the process.

This technique when appropriate is most effective for treating medium to high air flowrates with medium to high solvent concentrations. Similar disadvantages accrue as for fixed-bed adsorption. The airstream must be cooled before treatment and the costs associated with regeneration of the absorbent phase.

4. THERMAL INCINERATION - high temperature treatment of solvent contaminated air to chemically oxidise/ combust the organic compounds. The products of combustion or oxidation are carbon dioxide and water vapour and the reaction is usually exothermic. It is most important within an incinerator to achieve rapid and complete oxidation by operating at sufficiently high temperatures and avoiding quenching and channelling at the combustion chamber walls. Adequate turbulence within the combustion zone is also essential. Because the calorific value of the waste air is only typically of the order 1 MJ m^{-3}, a dual fuel fired system is necessary to ensure good combustion.

   The temperatures and residence time required depends upon the solvent composition. 90% destruction can be obtained with a combustion temperature of 870C and residence time of 0.75 seconds. At concentrations below 2000 ppm the efficiency reduces. For halogenated solvents, to achieve a 98% destruction efficiency requires a combustion temperature of about 1100C for 1 second. In this latter case the problems arising from hydrochloric acid emissions also need to be taken into account.

   Incineration is a relatively simple process and consequently widely used for clean-up of dryer exhaust streams. Its main disadvantage is the high energy requirement although modern designs incorporate heat recovery to improve the energy utilisation. Another potential problem because of the high temperatures employed are NOX emissions.

4. CATALYTIC INCINERATION - operates on a similar principle to thermal incineration except the combustion is carried out over supported metal catalysts which enables combustion to be carried out at much lower temperatures of the order of 250 to 450C and with ten to twenty times shorter residence times. Because of the lower temperatures and reduced residence times the equipment is much smaller than for thermal incineration and more economic. To initiate the combustion, preheating of the exhaust air is required but this can be simply achieved by using a heat recovery exchanger to recycle energy from the treated air.

Another advantage of catalytic incineration is that provided the solvent concentration is high enough, little or no fuel is required except for unit startup. This requirement will normally be fulfilled if high solvent loadings are achieved as shown in Table 3 below. The energy availability ratio is calculated assuming the ambient makeup air has to be heated on average to 200C dryer inlet temperature. The overall drying/ incineration process will be energy self-sufficient provided the ratio exceeds the value of two allowing for a 50% heat exchanger efficiency at maximum solvent loading of 50% LEL.

TABLE 3

Energy Self-Sufficiency of Solvent Laden Air for Catalytic Incineration

Solvent	Maximum Solvent Loading $/gg^{-1}$	Nett Heat Combustion of Solvent $/MJ\ kg^{-1}$	Energy Avail Ratio
dichloromethane	.0229	6.05	0.69
methanol	.0419	19.91	4.13
methyl chloroform	.1919	7.31	6.94
vinyl trichloride	.2069	7.23	7.40
ethanol	.0349	26.82	4.63
ethylene glycol	.0348	17.04	2.94
acetone	.0264	28.57	3.73
methyl acetate	.0403	18.99	3.79
isopropanol	.0210	30.45	3.17
n-propanol	.0210	31.47	3.27
vinyl acetate	.0391	22.64	4.38
methyl ethyl ketone	.0226	31.37	3.51
tetrahydrofuran	.0251	32.24	4.01
diethylene oxide	.0307	24.83	3.77
ethyl acetate	.0338	23.41	3.92
butanol	.0180	33.14	2.95
diethylene glycol	.0370	20.31	3.72
n-butylamine	.0216	37.96	6.95
propyl acetate	.0320	26.56	4.21
cyclohexanone	.0170	33.61	2.83
cyclohexane	.0195	43.44	4.19
isobutyl methyl ketone	.0244	34.82	4.21
n-butyl acetate	.0344	28.26	4.81
isobutyl acetate	.0487	29.19	7.04
diacetone alcohol	.0364	28.42	5.12
hexane	.0180	44.74	3.99
toluene	.0192	40.52	3.65
heptane	.0209	44.56	4.61
xylene	.0184	40.81	3.72

Note if the maximum solvent loading is not feasible then a proportionally higher energy availability ratio will be required for self-sufficiency. Similarly if higher or lower operating temperatures are required for the dryer then the ratio value will also need to be revised accordingly.

ETS CATALYTIC INCINERATOR

The above discussion has demonstrated the clear advantages of the catalytic combustion technique for treatment of solvent laden exhaust streams from dryers. One of the most interesting commercially available catalytic combustion systems are the units supplied by Eco Trend System (ETS), manufactured in Italy. The author's organisation is a UK representative and distributor for these catalytic incinerators.

The equipment has been assembled to provide a packaged unit which is suitable for easy connection to an existing system. It comprises centrifugal fan, filter system, air/air heat exchanger, catalytic combustion section, fuel system and appropriate controls. The whole unit is enclosed in a box structure with aluminium insulated panelling. The system has been designed with interchangeable panels so that the ductwork connections are feasible on all sides and with the ability that towers can be connected in series. The system units are of modular construction of capacity of 2500 Nm3 per hour. This represents a solvent evaporation rate of 35 to 100kg hr^{-1}, depending upon the maximum solvent loading of the airstream.

The unit operates as set out in the schematics in Figs. 1 and 2 which illustrate a typical system. The solvent laden air is sucked into the tower by a centrifugal fan through a heat exchanger (RC) where the temperature is raised. After passing through a series of filters to separate out unwanted liquid and solid particles, it enters the catalytic combustion chamber (CC) where the temperature of the solvent vapours raises to 350C.

The oxidation reaction takes place between the solvent and oxygen molecules as the air passes through the bed of nobel metal based catalyst particles to give an optimal thermal conversion of nearly 100% and solvent reduction of at least 75%. After the combuster, the air passes through a second-stage catalytic burner (FC), where the catalyst is supported on a ceramic honeycomb structure, to complete oxidation of the remaining vapours. Overall conversions of 90-99% can be achieved.

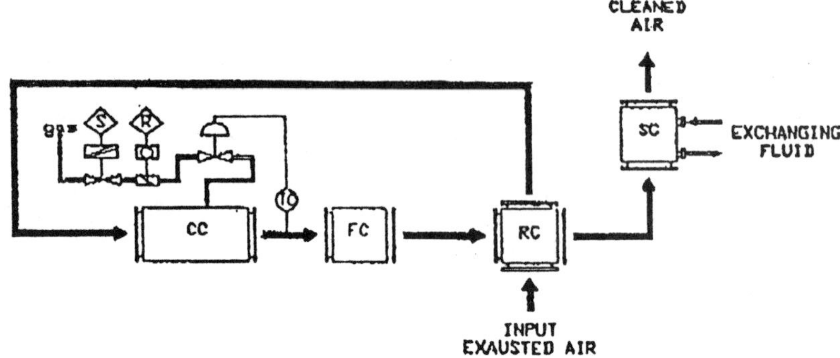

Legend:

GAS - combustible fluid	CC - catalytic burner
S - safety logic	FC - cold catalytic filter
R - control logic	RC - main heater exchanger
TC - temperature control	SC - sec. heater exchanger

FIG. 1 SCHEMATIC OUTLINE OF INCINERATION PROCESS

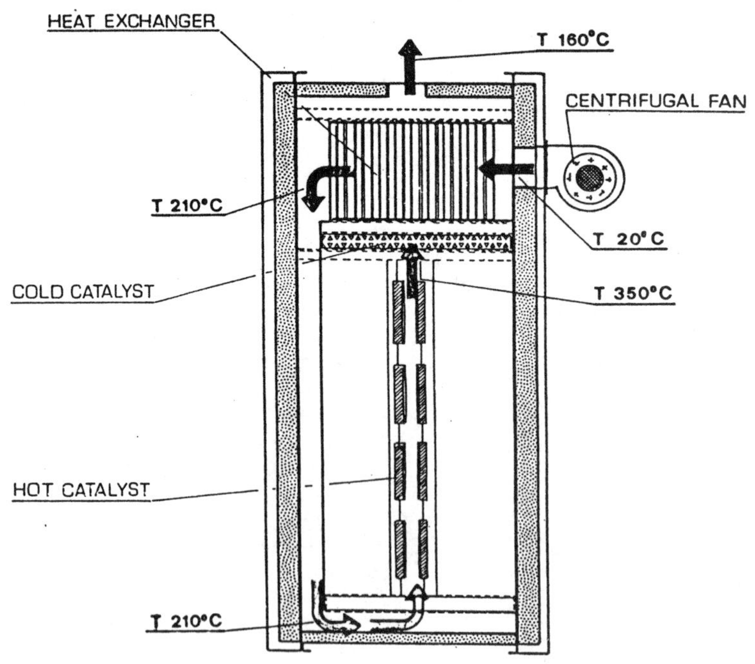

FIG. 2 DRAWING OF ETS "TD" UNIT SHOWING CATALYTIC COMBUSTION CHAMBER AND HEAT EXCHANGER SYSTEM.

The most suitable catalysts are alloys of platinum, palladium, rhodium and ruthenium coated on activated alumina. Such catalysts have an especially high stability and are resistant to most pollutants except lead, arsenic and phosphorus. The role of the catalyst in fume abatement is to increase the chemical rate of oxidation, thus allowing the reaction to proceed to equilibrium at a lower energy level. This is accomplished by diffusion of the solvent vapours from the bulk phase to the catalyst surface where chemisorption of the organic material to the active sites occurs. At this point, the oxidation reaction proceeds at an energy level dependent upon the orientation of the reactant in relation to the active site and the type and strength of the molecular bonds. After oxidation is complete, the products are desorbed from the catalyst surface and diffused back into the exhaust stream flow.

Care is taken in the design to ensure the correct catalytic ignition temperature and special turbulence promoters installed to ensure a minimum contact time. After the incineration is complete, the purified air is passed through the other side of the heat exchanger. Overall approximately 70% of the available energy can be recovered and used for heating purposes, drying, etc. As the energy required to achieve the ignition temperature is produced by the catalytic combustion process itself there is no need to supply additional energy from outside the system except for start-up purposes.

OVERALL FACTORS FOR DESIGN OF COMBINED SOLVENT DRYING AND TREATMENT SYSTEMS

The design of an overall dryer/treatment system for handling solvent based applications must inevitably involve a series of compromises to achieve the best performance. Factors which should be considered include:

1. NATURE AND QUANTITIES OF SOLVENT
   Can the coating technique be modified to enable more concentrated coatings to be applied to the substrate to reduce the quantity of solvent used ?
   What is the best solvent formulation to produce the required coating ?
   Are there other solvents which may be feasible which have more favourable processing properties (as defined by the MSL and PCI indexes) ?
   Can I avoid the use of solvents altogether and use a water-based formulation ?
   What effect do different solvent formulations have upon the drying properties of the coating ?

2. DRYER AIR CONFIGURATIONS
   Can the drver be reconfiqured to minimise the air usage to reduce treatment costs by maximisation of solvent loading ?
   Is it cost effective to switch to inert gas operation ?
   Examine the feasibility of installation of solvent monitoring instrumentation to enable the 50% LEL limit to be approached.

3. OPTIMISATION OF DRYER OPERATING CONDITIONS
   Examine interactions between dryer line speed, air temperatures, solvent loadings, and nozzle velocities. Are there sets of operating conditions, particularly at the feed end, which lead to too high evaporation rates leading to unsafe solvent concentrations ?

4. COST OF SOLVENT AND CHOICE OF TREATMENT TECHNIQUE
   As incineration means the solvent itself provides most of the dryer energy supply, is this an excessively expensive way of fuelling the plant compared to say the use of natural gas at 30p per therm ?
   Note in those circumstances the energy cost is directly related to the solvent cost and it's nett heat of combustion (see Table 3). For example, if the solvent costs £500 per tonne then the energy obtained from incineration represents a cost per energy unit of £1.19 to £3.10 per therm for non-chlorinated VOCs. For chlorinated VOCs, the equivalent cost is £7.23 to £8.73 per therm.
   Based on the above figures or in cases where the solvent is more expensive, is it cost effective in terms of extra capital investment and energy costs to use the direct solvent recovery techniques or combined solvent recovery/incineration systems ?

5. SCOPE FOR OVERALL SYSTEM CONFIGURATION
   Particularly for multizoned dryer systems, are there clever ways of linking zones together to minimise operating costs by use of integrated heat recovery exchanger/solvent treatment layouts ?

The above discussion provides a number of suggestions as to how the efficiency and cost effectiveness of solvent drying systems can be upgraded. However the dominant factors are the quantity of solvent used, flowrate of exhaust air, and the concentration of solvent in the exhaust air. It should also be realised that these three factors strongly interact with each other and have a dramatic effect upon the size, capital cost and operating cost of the air treatment system. This is best illustrated by the case-study in the following section.

CASE-STUDY OF ECONOMICS OF SOLVENT TREATMENT SYSTEMS

This industrial case-study example is based upon coating of plastic film with an ethyl acetate based adhesive used for manufacture of sellotape as described elsewhere (refs. 9 and 10). Typical operating parameters are as follows:

Evaporation rate of solvent = 140 kg hr^{-1}
Hours operation per year = 240 x 24 = 5760
Exhaust air temperature from dryer = 100C

MSL value for ethyl acetate = .0338 g g^{-1} (Table 1)

Minimum quantity of exhaust air = 140 / .0338
= 4142 kg hr^{-1}
= 3373 m3 hr^{-1} @ NTP

1. DESIGN OF THERMAL INCINERATOR SYSTEM

A modern incinerator system will comprise the process layout presented in Fig. 3, where a primary heat exchanger is used in conjunction with the incinerator to preheat the effluent airstream. It is possible to show the most economic design is to size the heat exchanger so that almost all the energy required for incineration is provided by the combustion of the solvent. If this state can be achieved, then the fuel consumption of the burner is minimised to just provide sufficient energy to initiate incineration.

The sizing of the heat exchanger is a direct function of the solvent loading of the airstream because of the interaction between the extent of preheating required and the temperature rise through the incinerator due to solvent combustion. That is, as the solvent level is increased the heat exchanger efficiency and size required is decreased. For example, for identical airflows the size of exchanger is doubled as the efficiency is uprated from 60% to 84%. Also the higher the solvent concentration the smaller the volume of air to be treated and the size of plant required, as shown in Table 4 below. For these design calculations it is assumed that the minimum inlet temperature of the airstream to the incinerator is 50C greater than the auto ignition temperature (480C), and, that the minimum outlet temperature from the incinerator is 760C to ensure complete combustion of solvent.

The results presented in Table 4 demonstrate the dramatic advantages of configuring the system so that the solvent concentrations in the airstream are maximised. It can be seen that the incinerator size decreases by a factor of eight and the heat exchanger size by a factor of twenty-one as the VOC concentration is enhanced from 5% to 30% LEL.

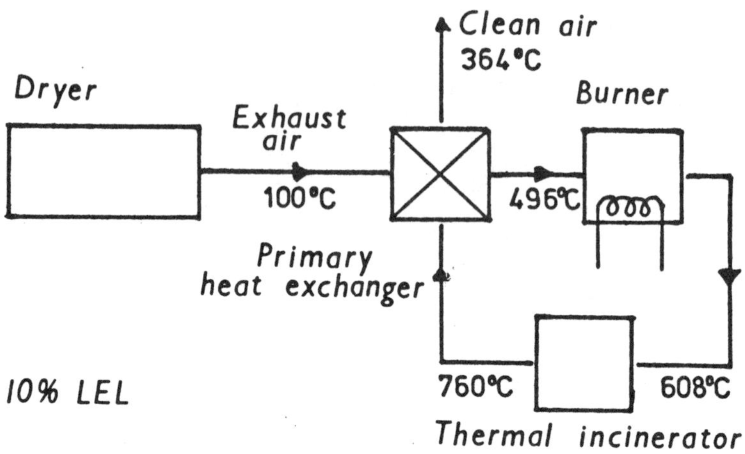

FIG. 3 SCHEMATIC LAYOUT OF THERMAL INCINERATOR SYSTEM

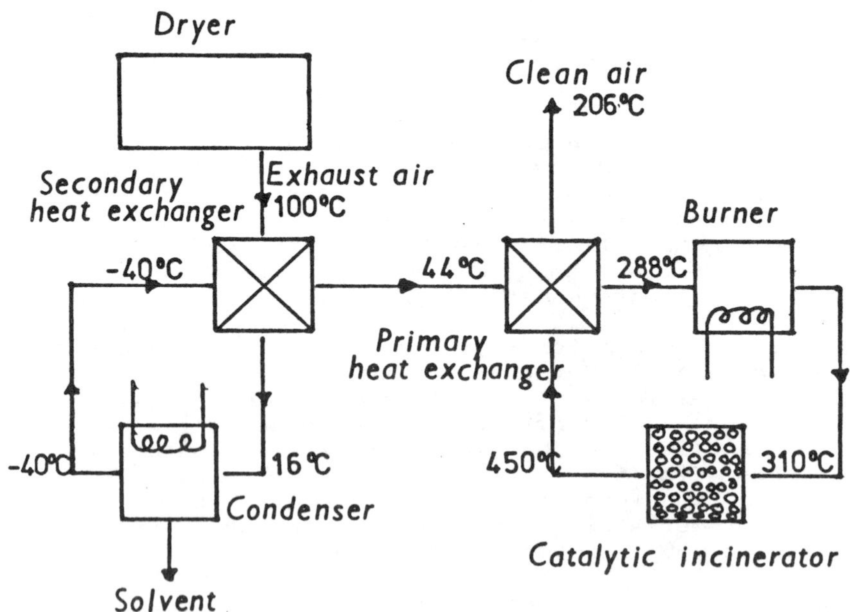

FIG. 4 SCHEMATIC LAYOUT OF COMBINED SOLVENT RECOVERY AND CATALYTIC INCINERATOR SYSTEM

TABLE 4

Effect of Airstream Solvent Concentration Upon Economics of Thermal Incinerator Design

Solv Conc /%LEL	Air Volume /Nm^3hr^{-1}	Incin Temp Rise /C	Heat Excha Effic /%	Heat Exch Size /-	Incin Size /-	Overall System Cost /£
5.0	33,730	76	88.5	21.25	8.43	452,000
7.5	22,487	114	82.7	11.48	5.62	306,000
10.0	16,865	152	77.0	7.23	4.22	230,000
12.5	13,492	190	71.2	4.88	3.37	184,000
15.0	11,243	228	65.5	3.48	2.81	153,000
17.5	9,637	266	59.7	2.56	2.07	131,000
20.0	8,433	304	55.5	1.99	1.75	112,000
25.0	6,746	380	50.0	1.36	1.30	87,000
30.0	5,622	456	45.5	1.00	1.00	70,000

2. DESIGN OF CATALYTIC INCINERATOR SYSTEM

The layout of the catalytic incinerator system as shown in Figs. 1 and 2 is similar to the thermal incinerator design where the solvent energy of combustion is used to preheat the effluent airstream and initiate incineration. In this instance the standard ETS modular units are used with approximate capacity of 2500 Nm^3 per hour each and heat exchanger efficiencies of 60%. Because of the constraint on exchanger efficiency it is necessary at low solvent concentrations to burn additional fuel to provide sufficient preheating. For ethyl acetate, the recommended catalytic incinerator inlet and outlet temperatures are 310C and 450C. The main design parameters for the treatment system are summarised on the next page in Table 5.

Comparison of Tables 4 and 5 reveals the advantages of catalytic incineration over thermal incineration. The more compact units and lower operating temperatures mean the capital cost of the catalytic system is about 70% of the thermal based systems. However when handling low solvent concentration airstreams (<7.5% LEL) there is little to choose between the two types when the total operating costs of capital and energy are assessed.

As the solvent concentration is increased the catalytic system becomes more and more favourable and energy self-sufficiency is achieved at 10% LEL. At higher LEL concentrations there is a surplus of energy for incineration. This means that there is increasing scope to utilise the energy in a different manner via a combined solvent recovery/catalytic incineration system as discussed in the next section.

TABLE 5

Effect of Airstream Solvent Concentration Upon Economics of Catalytic Incinerator Design

Solv Conc /%LEL	Air Volume /Nm^3hr	Incin Temp Rise /C	Fuel Energ Reqd /MW	Number Incin Module /-	Catal Incin Cost /£	Fuel Energy Cost /£yr^{-1}
5.0	33,730	76	0.77	13	304,000	45,000
7.5	22,487	114	0.27	9	204,000	16,000
10.0	16,865	152	-0.07	6	150,000	-4,000
12.5	13,492	190	-0.24	5	120,000	-14,000
15.0	11,243	228	-0.35	4	100,000	-21,000
17.5	9,637	266	-0.43	4	89,000	-25,000
20.0	8,433	304	-0.49	3	75,000	-29,000
25.0	6,746	380	-0.58	3	62,000	-34,000
30.0	5,622	456	-0.63	2	50,000	-37,000

3. DESIGN OF COMBINED SOLVENT RECOVERY AND CATALYTIC INCINERATOR SYSTEM

Although it has been shown in Table 5 that the catalytic incinerator system generates surplus energy when treating airstreams with high solvent concentrations, in practice this extra energy will be wasted. A much improved approach is to use a combined solvent recovery/ catalytic incineration system as sketched in Fig. 4. In this instance the effluent air from the dryer is partially chilled by a secondary heat exchanger to 16C (60% efficiency) and then passed through a refrigerated two-stage condenser. The first stage removes the bulk of the water present as no solvent will condense out above -16C, and, the second stage decreases the solvent level to 10% LEL by chilling to -40C. The chilled air is then preheated by the original effluent air via the secondary heat exchanger to 44C before final processing by the catalytic incinerator unit. Due to the remaining solvent level of 10%, the incineration is almost energy self-sufficient and only a small amount of preheating is required to initiate combustion.

The overall design parameters are summarised in Table 6 for the combined treatment system as a function of solvent concentration. The overall system cost includes the refrigeration plant, condensers, secondary heat exchanger and catalytic incinerator. The cost of the incinerator unit is the same as presented in Table 5. The energy costs take into account the 30% efficiency for the refrigeration plant using natural gas fuel at 30 pence per therm.

The value of solvent recovered is assessed at £400 per tonne to allow £100 per tonne purification costs. The overall operating cost sets off the amortisation cost of the equipment, based upon a 30% capital recovery factor per year, and the energy cost against the value of solvent recovered.

TABLE 6

Effect of Airstream Solvent Concentration Upon Economics of Solvent Recovery/Catalytic Incinerator System

Solv Conc /%LEL	Air Volume /Nm^3hr^{-1}	Refrig Energy Reqd /kW	Total Energ Reqd /kW	Total System Cost /£	Value Recov Solvent /$£yr^{-1}$	Overall Operat Cost /$£yr^{-1}$
15.0	11,243	286	1041	380,000	107,000	68,000
17.5	9,637	247	898	341,000	138,000	17,000
20.0	8,433	219	796	306,000	161,000	-22,000
25.0	6,746	178	646	261,000	193,000	-77,000
30.0	5,622	151	547	224,000	215,000	-115,000
35.0	4,819	132	478	204,000	230,000	-141,000
40.0	4,217	117	423	187,000	242,000	-160,000

Table 6 shows that the combined air treatment is very cost effective due to the cost savings from recovered solvent if the overall drying system can be configured to ensure suitable solvent levels in the effluent air. In effect if the system is properly designed then the air treatment system should prove self-financing. For example, if these operating cost savings are expressed instead in terms of payback, it can be shown that solvent concentrations of 23% LEL and 35% LEL offer two year and one year returns on the plant investment.

CONCLUSIONS

The results of the case study have demonstrated that the processing of VOC contaminated airstreams can be cost advantageous if the treatment system is carefully designed. For other applications depending upon the solvent properties, scale of operation, temperature of process exhaust air etc. other types of systems may be more appropriate using instead combined activated carbon adsorption/catalytic incinerator or liquid absorption/catalytic incinerator systems. It should also be noted that further cost savings may be possible by utilising the energy available in the treated air by tertiary heat recovery to assist space heating and hot water supplies.

The key to effective solvent handling is the maximisation of the VOC content in the exhaust air. Although in practice it is probably impossible to attain the maximum 50% LEL limit, the dryer system should be carefully configured to approach this limit as closely as possible. To achieve such results, detailed understanding of the performance of the drying system from zone to zone must be available. The information required can be provided by modelling software techniques developed by the author (ref. 11). It is the author's experience that even if the dryer is badly configured at present, there is usually a way by which the dryer configuration can be upgraded while preserving production capacity at modest cost.

The combined use of these modelling procedures together with the design procedures described earlier provides a powerful set of tools for design of pollution control systems. It should prove possible for most applications to provide cost effective designs based upon solvent recovery/catalytic incineration hardware which provide an excellent return on the capital investment. The modular nature of the catalytic incineration units also enables flexibility of operation needed for situations where the plant is used for multiproduct manufacture.

REFERENCES

1. S.T. Smith - 'New Pollution Control Framework', Paper presented at PITA Convertex '92.
2. HMIP - 'Process Guidance Note PC6/1, Printworks'. HMSO, 1992.
3. HMIP - 'Process Guidance Note PC6/10, Coating Manufacturing Processes', HMSO, 1992.
4. HMIP - 'Process Guidance Note PC6/12, Adhesive Coating Processes', HMSO, 1992.
5. HMIP - 'Process Guidance Note PC6/14, Film Coating Processes', HMSO, 1991.
6. HMIP - 'Process Guidance Note PC6/18, Paper Coating Processes', HMSO, 1992.
7. HSE - 'Occupational Exposure Limits', Guidance Note, EH40, HMSO, 1986.
8. J.G. Cleland and G.L. Kingsbury - 'Multimedia Environmental Goals for Environmental Assessment', Research Triangle Institute, EPA-600/7-77-136a.
9. B. Schofield - 'Waste Not Want Not, Pollution Control Equipment ', Paper presented at PITA Convertex '92.
10. D. Finch - 'Operating a Thermal Oxidiser', Paper presented at PITA Convertex '92.
11. J.C. Ashworth - 'New Computer Techniques for Optimisation of Drying of Coatings', Paper presented at Coating Technology Symposium, University of Bradford, Sept. 1992.

Design Considerations for Nitrogen Inerted Air Float Dryers

E.V. Bowden
SPOONER INDUSTRIES LIMITED, RAILWAY ROAD, ILKLEY,
WEST YORKSHIRE LS29 8JB, UK

1 INTRODUCTION

Solvent based coating formulations are still widely used in the converting industry despite the ever-increasing pressures imposed by anti-pollution legislation.

The use of solvents in coating formulation achieves product quality and performance criteria that cannot be achieved at present with water based formulations and so we can safely assume that processing with solvents will be a dominant factor in the coating industry for many years to come.

Ever-increasing energy and solvent costs, coupled with the environmental legislation pressures, have accelerated the development of new technologies for the recovery of volatile organic compounds (VOCs) from coating processes.

The inert gas condensation process is a significant development in this area, in that it provides a total solution to the pollution, solvent and energy cost problem.

Solvent recovery by inert gas condensation is already well established and has been applied successfully in a number of diverse coating operations.

A further development of the technique has been to combine it with high performance air flotation drying resulting in a total process system for today's sophisticated products which brings together all the advantages of the solvent recovery process with that of contactless, and hence scratch-free, processing.

2 EXAMPLES OF SUITABLE PROCESSES

Some examples of processes ideally suited to processing by inert gas condensation can be listed as follows.

- Magnetic Media (audio, video, floppy disc)
- Pressure Sensitive Tapes
- Coated Metals (steel, aluminium)
- Coated Films (electronic/packaging)
- Coated Textiles (printed circuit boards etc)
- Lacquered Foils (packaging, hot stamping)

Any process ideally suited to inert gas operation can usually be identified by applying the following criteria.

1. Is the solvent expensive?
2. Can the solvents be re-used in the coating mix?
3. Is drying carried out at relatively low temperatures?

The inert gas condensation process gains its economic advantage from two specific areas of operation:

- the cost of recovering solvent vapours from an inert atmosphere is probably the lowest of any available method
- the introduction of the inert gas into the dryer permits 100% recirculation of the drying atmosphere.

Collectively these two factors reduce the heat energy costs in the range 65% to 90%, depending upon the specific application.

3 CONVENTIONAL AIR DRYING

The traditional way of evaporating solvents from a coated layer is with jet impingement or contra-flow, forced convection air dryers.

Over the years the forced convection dryer has proved itself to be the most versatile system for obtaining consistent product quality coupled with high productivity rates and, moreover, the flexibility of the system permits different products to be processed on the same machine.

The typical air circuit of a jet impingement dryer operating with air is shown in Fig. 1. Even with solvent based processes the well designed air dryer will use high rates of recycle whenever possible. The running energy costs will be dictated by the quantity and temperature of exhaust air discharged from the process.

Safety legislation dictates that the vapour concentrations in a conventional air dryer must be kept safely away from explosive regions and typically 25% LEL is a selected operating level although, with suitable solvent concentration monitoring, it is possible to operate in the range 25 to 50% LEL.

The majority of solvents employed in coating operations (Fig. 2) will have lower explosive limits in the range 1% to 2% VV. The solvent concentration therefore in the conventionally operated air dryer is typically limited to 0.25 to 0.5% VV. In essence, large quantities of exhaust air are ever present and this in turn dictates high energy losses and inefficient operation.

Solvent recovery by condensation from the exhaust stream of such equipment is impractical due to the vapour pressure regime imposed by the low concentrations.

4 OPERATION IN INERT ATMOSPHERE

The limitations imposed by low solvent concentration are no longer applicable if the dryer atmosphere is replaced with an inert gas such as nitrogen (N_2). This is principally due to the fact that, in the absence of oxygen, the nitrogen alone cannot support combustion and thereby the presence of flammable or explosive gas mixtures within the oven circulation system is totally eliminated.

In the nitrogen inerted dryer the normal air environment, containing circa 23% oxygen, is replaced by nitrogen with the object of maintaining the oxygen content at 4% or less. Under these circumstances the concentration of solvent in the circulating system of the dryer can be safely raised to levels where the condensation process is viable and typically the system will operate in the range 5 to 40% by volume. (Fig. 3).

5 AIR FLOTATION DRYING

The nitrogen inerted drying process is ideally suited to work in conjunction with air flotation drying. This

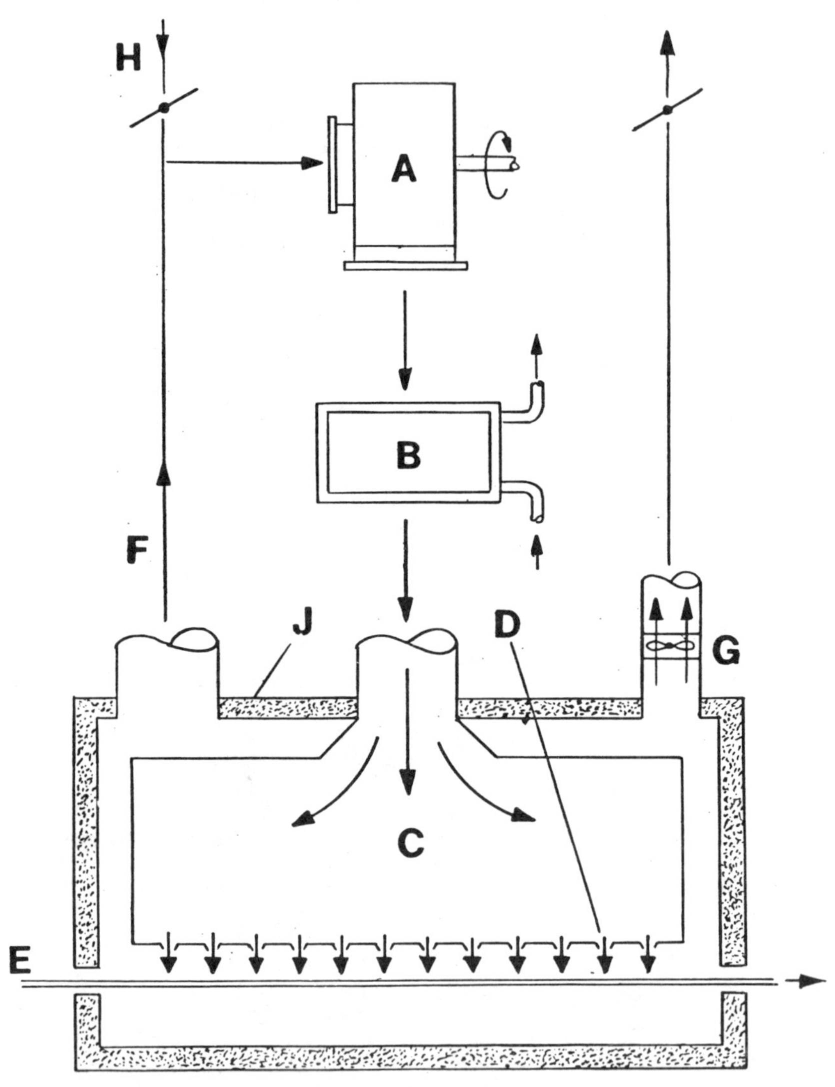

BASIC REQUIREMENTS OF A JET IMPINGEMENT DRYER

FIG 1

EXAMPLES OF RECOVERABLE SOLVENTS

Ketones

Acetone
Methyl Ethyl Ketone (MEK)
Methyl Isobutyl Ketone
Cyclohexanone

Ethers

Ethyl Ether
Tetrahydrofuran
Dioxane

Aromatics

Benzene
Toluene
Xylene
Napthalene

Hydrocarbons

Hexane
Cyclohexane
Heptane
Mineral Spirits

Alcohols

Methanol
Ethanol
Isopropanol

Chlorinated Hydrocarbons

Methylene Chloride
Methyl Chloroform
Perchloroethylene

Monomers

Vinyl Acetate
Acrylic Acid
Acrylonitrile

Esters

Vinyl Acetate
Ethyl Acetate
Isopropyl Acetate

FIG. 2

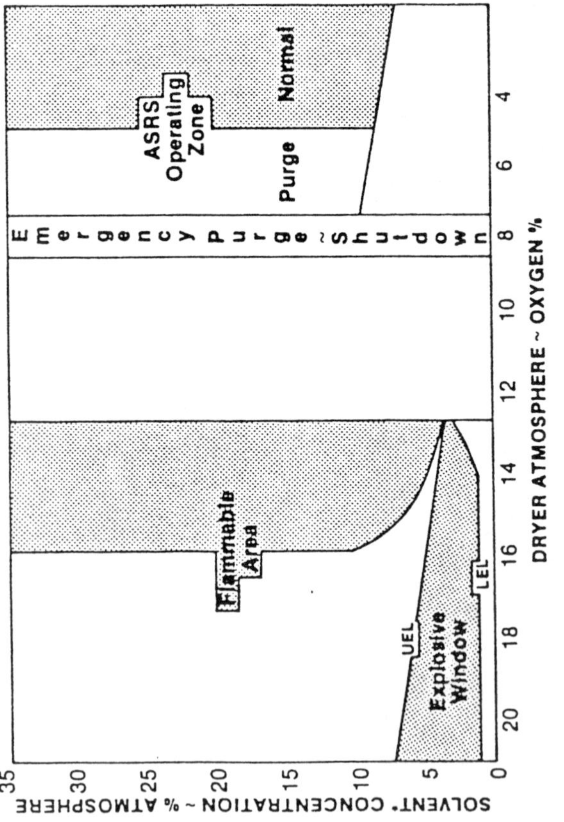

FIG. 2 — Control of Oxygen Concentration. *Source: CRC Handbook of Chemistry and Physics, Robert C. Weast, Ed., pg. D-106; CRC Press, Cleveland, Ohio 56th Edition.*

FIG. 3

technique ensures high product quality together with contactless operation throughout the whole of the drying process.

A system such as "Spooner Float" permits the full use of all the process variables available with the inert gas system. This is true for both one sided coated processes as well as those requiring coatings applied simultaneously to both sides, these include impregnated products where the base substrate is totally saturated by immersion in the coating solution.

It is vital to ensure that any air flotation system under consideration will meet the exacting requirements demanded of it by the often exacting process and product requirements.

The characteristics of operation of the high performance "Spooner Float" system are as listed here.

- Complete Web Stability
- Wide Range of Substrates
- Wide Operating Tension Range
- Sine Wave Generation
- Reynolds Number Effect
- High Operating Clearance

6 WEB STABILITY

Of utmost importance with any air flotation system is that the web passing through the process should be free of any tendency to flutter or vibrate under the influence of the impinging air velocity, to ensure contamination-free operation ensues under all operating conditions.

Modern air flotation systems such as the "Spooner Float" Air Bar will operate over a wide range of process variables whilst still maintaining absolute web stability. A first class air bar will permit the use of 10 to 100 metres/second at the slot outlets without any loss of air flotation performance.

Support of the product is provided by generating a pressure pad between the surface of the air bar and the web being processed (Fig. 4). In the "Spooner Float" air bar this is generated by using unique aerodynamic effects at the surface,brought about by accelerating air streams giving high and low pressure regions (Coanda effect). These result in a stable air pressure cushion of high magnitude.

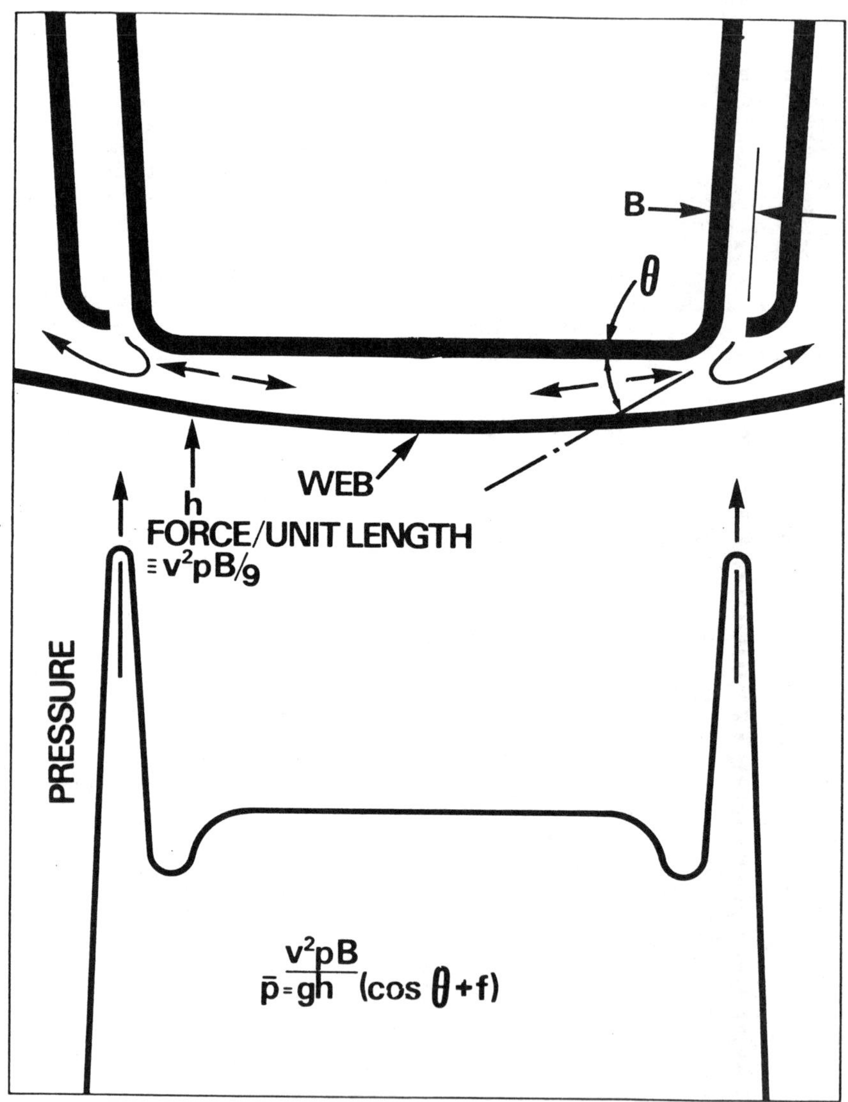

Fig. 4

Spooner - Sine wave configuration
Exhaust Area
Impingement Areas
Boundary Layer
Coating
Substrate

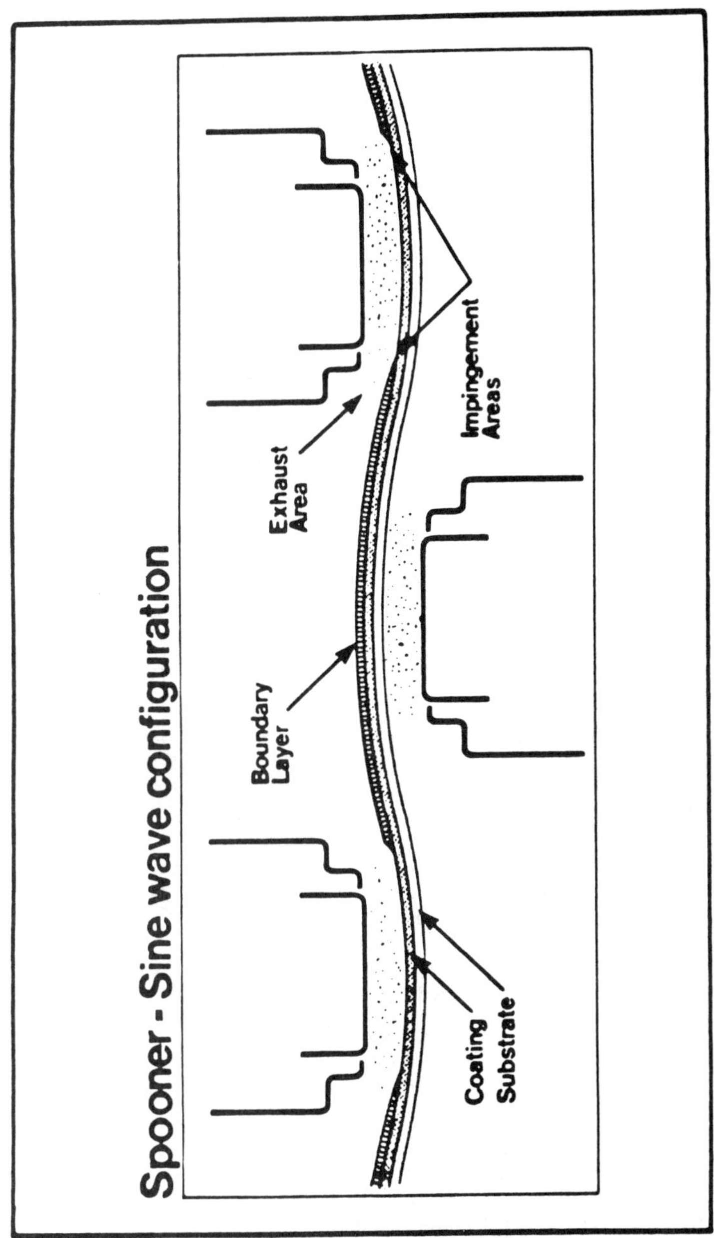

Fig. 5

By placing one air bar above the other in a stagger fashion the air bar system can be made to induce a sinusoidal wave form into the web against the prevailing web tension. If sufficiently high in amplitude the sine wave form imparts a beam stiffness to the product, which in turn helps to suppress the formation of creases and edge curl during the drying operation. This characteristic is particularly important when processing lightweight coated paper webs and thin filmic substrates.

Fig. 5 shows an Air Flotation Dryer running on a paper based product at 500 metres/minute with an air velocity of 50 metres/second. The normal clearance between the product and the air bar surface is not less than 8 mm.

7 HEAT AND MASS TRANSFER

Stable air floation with good clearances is the hallmark of a good system but attendant with this must be the capacity to bring about high drying rates, since the performance of any drying system is measured in terms of its ability to remove moisture or solvent to the desired retention level, always providing this is consistent with good product quality.

Forced convection systems are unique in that good drying rates are not dependent upon the use of high process temperatures. This is particularly the case with air flotation dryers and the answer to this can be found in the Nusselt equation (Fig. 6) where it can be shown that the heat transfer rate, and hence the drying rate capability (Nu number), is predominantly a function of the Reynolds number (Re) and its dependency on the Reynolds number exponent (N). In simple terms this means that the drying rate of an air floation system can be varied widely and regulated precisely by changing the air velocity parameter and leaving the temperature constant.

NUSSELT EQUATION $NU = aRe^{\pi}oPr^{1/3}$ Fig. 6

In this way high drying rates are achievable without the use of high air temperatures and, if correctly designed, this is achieved without the loss of web stability or air flotation clearance.

The Reynolds number effect is particularly important when evaporating highly volatile solvents, where precise regulation of the air velocity can optimise the drying rate without loss of product quality.

8 OPERATION WITH NITROGEN (N_2)

In any drying system the mass transfer rate is regulated as a function of the available heat transfer coefficient and the prevailing vapour pressure difference between the circulating gas stream and that present at the product surface.

When operating with nitrogen containing high concentrations of solvent vapour, the available heat and mass transfer rates are somewhat different to that prevailing in the conventional air operated dryer not withstanding the same temperature and air bar velocity parameters.

Corrections to the heat transfer coefficient must be made in accordance with the variations in the gas transport properties. This variable tends to be small however, when compared to changes in the vapour pressure difference which is, after all, the driving force for the mass transfer operation.

Clearly at high solvent concentrations the vapour pressure difference is greatly reduced and this introduces a third drying parameter which can be used effectively to suppress the evaporation of solvent in the early stages of the drying process.

The use of the third parameter is particularly effective when considering processes involving highly volatile solvents prone to causing product quality problems due to the rate of solvent release in the early stages of drying, this phenomena is difficult to suppress within the normal parameters of an air drying system.

By careful computer modelling the nitrogen inerted drying process can be controlled, stage by stage, to ensure optimum drying rate during the sensitive stages of solvent evaporation, whilst still maintaining the requisite solvent retention levels on exit from the drying process.

9 NITROGEN INERTED SOLVENT RECOVERY SYSTEM

The nitrogen inerted air flotation dryer in circuit with the solvent recovery system is shown schematically in (Fig. 7).

The air flotation dryer is often multi-zoned for purposes of scheduling the gas velocities and temperatures in accordance with the best evaporation profile. Each zone has its own circulating fan system and heat source, using an indirect heat exchanger, usually heated by steam, hot oil or high pressure hot water.

COOLING MACHINE
LIN
LOW-TEMPERATURE CONDENSER
A
B
④
MAIN CONDENSER
②
③
HEAT EXCHANGER
①
SOLVENT SEPARATOR
FAN
INJECTION PUMP
INERT GAS DRYER

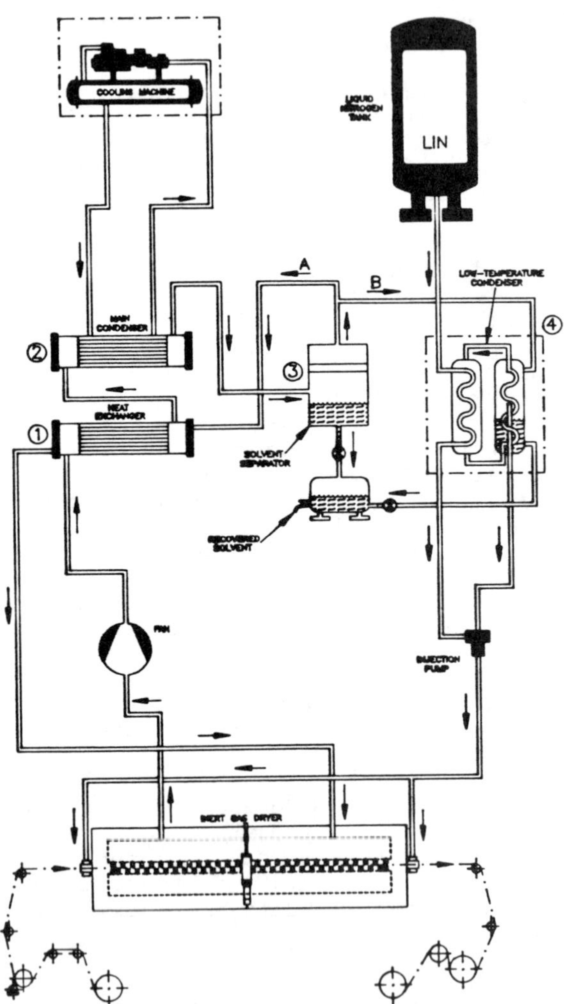

Fig. 7

The air velocity at the impingement air bars is best regulated by an inverter controlled AC motor drive, directly mounted on the circulating fan. This enables precise air velocity control to be achieved over a very wide working range. Alternatively, velocity control can be achieved by flow dampers but careful attention to damper design is required.

A small quantity of solvent-laden nitrogen is extracted from zone 1 of the dryer. At this point the maximum concentration of solvent vapour is present in the system. The quantity of gas extracted is determined by the quantity of solvent being evaporated per unit time and the desired maximum solvent concentration.

The temperature and concentration of the exhaust stream is matched carefully to avoid the saturation temperature, this ensures that condensation cannot occur within the circulation system of the drying zones.

The solvent rich nitrogen stream is extracted via a blower unit and delivered to the condensation system. A series of heat exchangers is utilised to progressively drop the temperature in the gas stream to promote virtually total condensation. The liquid solvent is collected in a sump at the base of the heat exchanger unit where it is eventually delivered to a solvent collection vessel.

The final stage of cooling is via an indirect heat exchanger which is cooled to very low temperatures by using the liquid nitrogen supply, also used for initial inerting of the dryer circuit. The extremely low temperatures available in the final stage (circa minus 195°C) ensure maximum removal of solvent from the circulating nitrogen stream. The stripped nitrogen is then returned to the dryer and re-introduced into the circulating system of the final dryer zone.

Low levels of solvent retention in the dried product are maintained by ensuring that the final zone of the dryer is fed with the relatively clean nitrogen returning back from the condensation unit and also by virtue of the fact that, at this stage in the drying operation, very little solvent is removed from the product by virtue of the evaporation profile, such that the circulating concentration at this point in the operation is kept relatively low.

The circulating nitrogen quantity is then cascaded through the zones of the dryer from the last zone to the first zone by an internal transfer system and in doing so the solvent concentration is progressively raised in accordance with the evaporation profile.

The relatively small quantity of clean fresh nitrogen liberated by the final stage cooling is then delivered to the ends of the drying system, where it is used to assist in sealing the entry and exit web slots to avoid the ingress of oxygen and/or expulsion of solvent laden nitrogen.

10 DRYER DESIGN CONSIDERATIONS

The design of an air flotation dryer suitable for inert operation with nitrogen is quite different to a unit operating with air as the drying medium.

The most important consideration, of course, is that of totally sealing the dryer to prevent the ingress of oxygen into the circulating gas stream.

Very special construction techniques and materials resistant to corrosion and solvent attack must be used throughout and much expertise is needed to avoid the many pitfalls associated with a design of this type.

The internal volume of the dryer should be kept as small as possible to avoid the unnecessary use of costly nitrogen during the initial filling operation (purging).

The outer skin of the dryer enclosure must be designed to facilitate total seam welding and on completion it is essential to pressure test for leakage.

In essence pressure testing consists of sealing up the entry and exit web slots then pressurising the system to evaluate the decay rate against time, and alternatively the leakage flow rate present. This must meet acceptable criteria before the drying system can be delivered to the end customer.

11 INTERNAL COMPONENTS

The well designed inert dryer will house the circulating fan and heater system inside the dryer body enclosure. Very often absolute filters are also placed in circuit inside the dryer where ultra-clean processes such as magnetic media products are under consideration. All of these internal components need to be accessed for maintenance purposes and the best way to achieve this is to design the fan, heater and filter units to be withdrawable from the rear of the drying enclosure. All components of this type must be purposely designed for flange sealing and one piece gaskets made from a suitable solvent proof material must be employed.

Of particular importance is the circulating fan and dynamic sealing fan, these need to be designed so that the motor bearings and fan rotor can be drawn from the fan case, which in turn is welded and sealed into the dryer body.

In addition the fan must incorporate a fan shaft seal which prevents the ingress of oxygen or, alternatively, the loss of solvent laden nitrogen from within the pressurised fan case. Shaft seals are available utilising mechanical or gaseous means of sealing. Of most interest perhaps is a particular sealing technique which uses a small flow of fresh nitrogen to effect a gas barrier around the rotating fan shaft.

Particular attention must be paid to any pipes, services or measuring points which must pass through the outer skin of the enclosure, since these are potential areas of leakage and solvent condensation.

Inerted air flotation dryers are often of an overall length which prohibits installation in one piece at the customer's premises and the design must allow for shipping in discreet pieces, each of which is completed and fully seal tested.

The design must also facilitate the joining and seam welding of each dryer zone, one to another, in situ ensuring alignment and outer skin integrity.

12 ENTRY/EXIT SEALS

Of particular importance in any inerted dryer is the provision of suitable seals where the web enters and leaves the drying enclosure. This area in itself presents the biggest potential leakage point and poses the most difficult design problems.

In a flotation dryer the end seal design problems are exacerbated by the fact that the contactless mode of operation on both sides of the web, must be observed in these areas, as well as throughout the full drying length.

The seal design employed by Spooner Industries is the result of a prolonged research and development programme during which many different approaches to the sealing problem were investigated.

Previous experience had used labyrinth sealed designs on the coated side and contact roller seals on the under side of the web being processed. This design of seal is only applicable to roller supported dryers utilising non-air flotation nozzles on the coated side of the product only.

The air flotation gas seal marketed by Spooner Industries Ltd. consists of a three-stage seal system the first and second stages are formed by dynamic sealing techiques and the third is of the labyrinth design.

The dynamic seals use fully and/or partially contaminated nitrogen driven by variable gas velocities which present angled jet curtains, these work against any internal turbulence and inner enclosure pressure, to prevent loss of contaminated nitrogen from within the drying enclosure even when the dryer body pressure is greater than the incumbent atmospheric pressure outside of the drying unit.

The third or labyrinth seal is fed with fresh nitrogen from the final stage of the evaporation process. It is convenient to use the fresh gas liberated in this way into the seals, although in truth this quantity, whilst kept to a minimum for cost reasons, is in theory extraneous to the mass balancing requirements of the dryer if correctly designed and sealed.

The labyrinth section of the gas seal floods the area immediately above and below the web through a serious of vanes and this ensures that any gas movement into or out of the dryer system, caused by pressure imbalances, will induce fresh nitrogen to enter or leave the seal section so that contaminated nitrogen can never enter the machine house or, alternatively, oxygen enter the circulating system of the dryer unit.

Fresh nitrogen is in itself expensive, and the main objective in designing a gas seal is that it should not, in itself, demand the use of nitrogen to make it operate. In effect the dynamic section of the Spooner gas seal, which utilises the circulating contaminated nitrogen, will in itself perform an adequate sealing duty and the fresh nitrogen requirement is utilised as an extra precautionary measure.

13 DRYER ACCESS

The nitrogen inerting process poses a conundrum to the dryer designer in that perfect sealing of the enclosure must be achieved whilst at the same time maintaining good access to the internals of the dryer for cleaning and maintenance etc. A further consideration is that in the absence of a special purpose web threading device sufficient access must be provided to permit the operator to feed up the web through the whole length of the dryer enclosure by hand.

Access doors and panels with sealing gaskets have all been used in designs in the past, with varying degrees of success. Even designs permitting the operator to walk inside the unit have been proposed but these, of course, present a potential safety hazard if pockets of nitrogen can remain undetected inside the dryer body after the system has been purged and air introduced into the circuit.

The total problem posed by providing adequate access with 100% sealing and interlocking precludes the use of doors and access panels for practical reasons.

Spooner Industries' solution to the problem has been to provide a design which permits the upper half of the dryer to be raised via a front side actuator on a special rear side hinge system. This separates the upper and lower air bars by as much as 600 mm permitting full access to the total length of the drying enclosure. Much design expertise has gone into making this form of retraction a reality, not least of which, of course, is effective peripheral sealing at the joint between the upper and lower halves.

Many factors must be taken into account when designing a practical joint of this type, of major importance is the joint material itself and the way it is mounted and supported in use. A further complication is that the joint must be easily removed and replaced as a one piece unit should inadvertent damage occur, or as and when seal maintenance is required.

14 CONCLUSION

The nitrogen inerted condensation process presents many challenges to the dryer manufacturer.

Conventional dryer models do not accurately describe the process and new models must be evolved. The process also presents unique problems to the dryer designer who, in turn, will rely upon research and development projects to provide the answers to difficult sealing problems before a working design can be evolved.

Additionally the technique involves total system knowledge, as well as an in-depth understanding of the processes being carried out and consequently few dryer manufacturers are capable of offering the one-source expertise necessary to put together a complete system.

The capital investment required for the equipment is inevitably high but the potential rewards are great in terms of total elimination of emissions to atmosphere, huge reductions in energy usage, together with the added bonus of virtually total recovery of evaporated solvents for process re-use.

All in all, the nitrogen inerted solvent condensation process offers an elegant solution to the problems faced by today's convertors engaged in solvent based coating operations.

E.V. BOWDEN

Subject Index